Place, Migration and Development in the Third World

Routledge series on Geography and Environment
Edited by Michael Bradford

Urban Housing Provision and the Development Process
David Drakakis-Smith

David Harvey's Geography
John L. Paterson

Planning in the Soviet Union
Judith Pallot and Denis J.B. Shaw

Catastrophe Theory and Bifurcation
A.G. Wilson

Regional Landscapes and Humanistic Geography
Edward Relph

Crime and Environment
R.N. Davidson

Human Migration
G.J. Lewis

The Geography of Multinationals
Edited by Michael Taylor and Nigel Thrift

Urbanisation and Planning in the Third World: Spatial Perceptions and Public Participation
Robert B. Potter

Office Development: A Geographical Analysis
Michael Bateman

Urban Geography
David Clark

Retail and Commercial Planning
R.L. Davies

Institutions and Geographical Patterns
Edited by Robin Flowerdew

Uneven Development and Regionalism
Costis Hadjimichalis

Managing the City: The Aims and Impacts of Urban Policy
Edited by Brian Robson

International Geopolitical Analysis
Edited and translated by Pascal Giret and Eleonore Kofman

Analytical Behavioural Geography
R.G. Golledge and R.J. Stimson

Money and Votes: Constituency Campaign Spending and Election Results
R.J. Johnston

The Uncertain Future of the Urban Core
Edited by Christopher M. Law

Mathematical Programming Methods for Geographers and Planners
James Killen

The Land Problem in the Developed Economy
Andrew H. Dawson

Geography Since the Second World War
Edited by R.J. Johnston and P. Claval

The Geography of Western Europe
Paul L. Knox

The Geography of Underdevelopment
Dean Forbes

Regional Restructuring Under Advanced Capitalism
Edited by Phil O'Keefe

Multinationals and the Restructuring of the World Economy
Michael Taylor and Nigel Thrift

The Spatial Organisation of Corporations
Ian M. Clarke

The Geography of English Politics
R.J. Johnston

Women Attached: The Daily Lives of Women with Young Children
Jacqueline Tivers

The Geography of Health Services in Britain
Robin Haynes

Politics, Geography and Social Stratification
Edited by Keith Hoggart and Eleonore Kofman

Planning in Eastern Europe
Andrew H. Dawson

Planning Control: Philosophies, Prospects and Practice
Edited by M.L. Harrison and R. Murdey

Education and Society: Studies in the Politics, Sociology and Geography of Education
Edited by L. Bondi and M.H. Matthews

Place, Migration and Development in the Third World

An Alternative View

With particular reference to population
movements, labor market experiences and regional
change in Latin America

Lawrence A. Brown

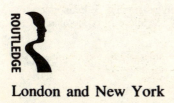

London and New York

First published 1991
by Routledge
11 New Fetter Lane, London EC4P 4EE

Simultaneously published in the USA and Canada
by Routledge
a division of Routledge, Chapman and Hall, Inc.
29 West 35th Street, New York, NY 10001

Printed and bound in Great Britain by
Biddles Ltd, Guildford and King's Lynn

British Library Cataloguing in Publication Data

Brown, Lawrence A.
  Place, migration, and development in the Third World:
  an alternative view – (Geography and the environment).
  1. Developing countries. Economic development. Latin America
    I. Title II. Series
    330.91724

    ISBN 0-415-05337-4

Library of Congress Cataloging in Publication Data

Brown, Lawrence A., 1935–
  Place, migration, and development in the Third World : an
alternative view / Lawrence A. Brown.
      p. cm. — (Routledge series on geography and environment)
  Includes bibliographical references.
  ISBN (invalid) 0-415-05337-4
  1. Developing countries. 2. Population geography. 3. Economic
development. 4. Regional economics. 5. Space in economics.
  I. Title. II.Series.
  HC59.7.B689   1990
  338.9′009172′4—dc20   90-32406
                                                   CIP

Contents

Tables   xi
Figures   xiii
Preface   xvii

1 Introduction   1

Development effects on population movements
in Third World settings   2

Rethinking Third World development   3

The odyssey of research and related features
of the study   6

2 What is Third World development?   10

A synopsis of development paradigms   10

Stage formulations   11
Dual society models   13
Human resource approaches   15
Political economy perspectives   17
The Latin American school of development   21

An alternative perspective: Third World development
as the local articulation of world economic and
political conditions, donor-nation actions, and
government policies   24

Paradigmatic statements concerning development:
a precis   26
Paradigmatic frameworks and Third World policies   29
Third World landscapes at ground level   30
Moving beyond paradigmatic thinking   34

Development or modernization?   37

A note on research strategy   39

3   **Aggregate migration flows and development,
     with a Costa Rican example**   41

Conventional modeling of migration in Third World
settings   42

Development–migration relationships in previous research   44

A development paradigm of migration   47

Intercantonal migration in Costa Rica and development
milieu effects   54

   Modeling procedures   55
   Substantive interpretation and expected relationships   58
   Empirical findings: application of the
      conventional model   59
   Empirical findings: spatially varying parameters of the
      conventional model   62
   Discussion   72

Alternative approaches to studying migration
in the aggregate   74

Summary and concluding observations   77

4   **Individual migration and place characteristics
     related to development in Venezuela**   79

Development context and individual migrations   80

   Background   81
   Indices of development for sub-national units
      of Venezuela   83
   Model specification and data characteristics   87
   Direct effects   91
   Interaction effects   93

Summary and concluding observations   96

Appendix 4.1: Development characteristics of Venezuela
                     distritos, 1971   100

5 Individual labor market experiences and place
   characteristics related to development in Venezuela 106

Labor market experiences, location, and social category 107

   Average labor market experiences and Venezuela's
      space-economy   108
   Research design focussing on individual labor
      market experiences   112
   Statistical analyses: educational attainment   114
   Statistical analyses: labor force participation   117
   Statistical analyses: wages received   119

Concluding observations   123

   Individual labor market experiences:
      a synopsis of findings   123
   General themes   124

Appendix 5.1: Mean values for variables employed
                     in regression analyses of individual
                     labor market experiences   127

6 Policy aspects of development and regional change I:
   Population movements from Ecuador's rural Sierra   128

Ecuador and its economy   129

Agrarian structure, land reform, and movement from the
rural Sierra: a broad-gauged portrait   131

   Agrarian structure   132
   Land reform and its impacts   134

Statistical analyses   136

Indices of structural conditions in Ecuador cantones   136

    Indices of agrarian structure   136
    Indices of the broader socioeconomic environment   138
    Articulation of structural indices in the
        Sierran setting   140

Linking place indices and variables representing
personal attributes   143

Findings   144

    Circulation   146
    Migration   149
    Land reform and movement from the rural Sierra   150

Broadening the perspective   151

Appendix 6.1: Provinces, cantones, and selected urban
                areas of Ecuador   155

7   Policy aspects of development and regional change II:
    The juxtaposition of national policies and local
    socioeconomic structures in Ecuador   157

Background   157

Regional change in Ecuador, 1974–82   159

    Principal components   160
    Canton groupings   163

National policies influencing regional change in Ecuador 166

    Price policies   167
    Credit policies   169
    Monetary exchange rate policies   170

National policies and local growth outcomes   172

    Variables representing structural conditions   172

Statistical analyses   176

Summary and concluding observations   180

Appendix 7.1: Component scores indicating socio-
                        economic change for Ecuador
                        cantones, 1974–82   182

Appendix 7.2: Ecuador cantones grouped according to
                        socioeconomic change profiles for the
                        period 1974–82, and related
                        characteristics   186

8 **Third World development as the local articulation
   of world economic and political conditions,
   donor-nation actions, and government
   policies: Concluding observations**   190

Considerations related to understanding
development in Third World settings   192

    The statistical representation of development and
        place knowledge   193
    The issue of regional change and research protocols
        for studying it   194

Summary observations and research implications   200

Notes   207

References   223

Index   246

Tables

3.1 Conventional models of migration: statistical estimates
for Costa Rica, 1968–73   60
3.2 Spatially varying parameter models of migration:
statistical estimates for Costa Rica, 1968–73   63
3.3 Regression-derived equations employed for mapping
spatially varying parameters (SVPs)   64

4.1 Principal components analysis of place variables for
178 Venezuela distritos, varimax rotation   85
4.2 Mean values of variables related to individual
out-migration in Venezuela   90
4.3 Logistic regressions for individual out-migration
in Venezuela, direct effects only   92
4.4 Logistic regressions for individual out-migration
in Venezuela, direct and interaction effects   95
4.5 Net, or summed, b-coefficients related to individual
out-migration in Venezuela   96

5.1 Regression statistics accounting for individual
educational attainment in Venezuela   116
5.2 Regression statistics accounting for individual
labor force participation in Venezuela   118
5.3 Average monthly wages (bolivares) among branches
of economic activity and employment patterns
for females and males, Venezuela   120
5.4 Regression statistics accounting for individual wages
in Venezuela   121

6.1 Principal components analysis of agrarian variables
for Ecuador cantones in 1974, varimax rotation   138
6.2 Principal components analysis of socioeconomic
variables for Ecuador cantones in 1974,
varimax rotation   139
6.3 Indices pertaining to agrarian and socioeconomic
structure, out-circulation, and out-migration
from the rural Sierra of Ecuador   141
6.4 Mean values and logistic regressions related to
out-circulation and out-migration from the
rural Sierra of Ecuador   145

7.1   Principal components analysis of socioeconomic
      variables representing change among Ecuador
      cantones for 1974–82, varimax rotation   161
7.2   Average principal component scores for canton
      groups, indicating the nature of socioeconomic
      change from 1974 through 1982   164
7.3   Means values for variables representing structural
      conditions of canton groups in 1974   175
7.4   Stepwise discriminant analysis of regional center,
      commercial agriculture, and domestic agriculture
      canton groups on the basis of variables
      representing structural conditions in 1974   178

Figures

2.1 Development stages as seen by Rostow (1960)   12
2.2 Core and periphery distinctions   15
2.3 Impact of colonization on the urban system
    of West Africa   19
2.4 Ideal–typical sequence of transportation development
    in a developing country   20

3.1 Zelinsky's hypothesis of the mobility transition   51
3.2 Shifts in the role of migration factors over the
    course of development   53
3.3 Costa Rica base map   65
3.4 Spatially varying parameters for out-migration:
    population pressure and percent of urban-based
    jobs at the origin   66
3.5 Spatially varying parameters for relocation: distance
    between origin and destination and population
    at the destination   68
3.6 Spatially varying parameters for relocation: percent
    of urban-based jobs at the destination   69
3.7 Spatially varying parameters for relocation: population
    pressure at the destination   70
3.8 Spatially varying parameters for relocation: average
    monthly wage per capita at the destination   71

4.1 Spatial distribution of development in Venezuela
    1971, and selected places   88

5.1 Labor market experiences and development in
    Venezuela, 1971   110
5.2 The spatial distribution of labor market experiences
    in Venezuela, 1971   111

6.1 Population movements from the rural Sierra   148

7.1 Spatial distribution of canton groups in Ecuador   165
7.2 Structural conditions mediating policy impacts
    among cantones   174

to

– JAFFE –

'a good man'

Preface

In addressing place, population movements, and development in Latin America and other Third World settings, this book intertwines themes that have personal as well as professional significance. A brief account of these underpinnings provides a frame of reference for the study itself.

My first contact with Latin America was in 1957. Two boys on the road, looking for a final adventure in the southwestern United States, dropped down to northern Mexico. I was especially captivated by Saltillo, perhaps because it was so exotic, so foreign, yet so close to my own country. Next summer, driving the Pan American Highway to Panama, I felt the rhythms of Hispanic culture; explored ruins of ancient civilizations; mixed with traditional Indian societies in Mexico and Guatemala; experienced (what then seemed to be) frontiers of civilization; encountered adventuresome hobos like myself, but from England, Germany, France, and Costa Rica, exotic places in their own right. One day, on a second-class bus in the Guatemalan highlands, my slouching drowse was interrupted by a clearly whistled classical refrain (Bach? Haydn?) floating over the multitude of Indians, and an international development person strode into view. Among the summer's many events, this stands as a symbol of all that touched me. Three years later, after a stint in accounting and law but having decided I wanted to do something 'meaningful' (in early 1960s terms), I became a geographer.

Another long-standing concern is development. In becoming a geographer, I was motivated by the prospect of Third World change wherein multitudes would live a better life and tomorrow would represent hope, opportunity, and personal growth. This was an achievable, realistic goal in the halcyon naivete of the early 1960s. Kennedy assured us it could be attained, the Civil Rights movement demonstrated it, and the Peace Corps embodied it. Twenty-five years of experience has tarnished the ideal, and I've become skeptical concerning broad-based development at the institutional level. Remaining, however, is a belief in beneficial change of a limited and incremental nature.

Development also is a personal concept; the idea of an individual growing, being satisfied rather than frustrated, realizing his/her full potential, self-actualizing in the Maslow (1968) sense, and the like. Becoming an educator represents a commitment to personal development, one that has grown and become increasingly important through the years.

My 'sense of place' is another element of this book. I've spent enormous amounts of time exploring both rural and urban areas, feeling the texture of landscapes, noticing who lives where and what they do, talking with strangers, crawling through factories, barns, and ships. The laboratory of these endeavors includes exotica such as southeast Ohio, Cedar Rapids, Iowa, and Philadelphia, Pennsylvania, as well as foreign areas. Place curiosity started as a hobby, but became a professional tool.

In June 1977, on Rio de Janeiro's Copacabana beach, I related much of the above to Harold Wood, then a McMaster University professor and President of the Geography Commission of the Pan American Institute of Geography and History. Despite my motivation for becoming a geographer, research in Latin America had never materialized, and feeling vulnerable to questions concerning sincerity, I broached the subject with caution and trepidation. That was unnecessary. As an educator with commitment to personal development as well as Latin America, Harold jumped at my hint, offering assistance whenever needed. His knowledge of Latin American institutions, wide range of personal acquaintances, unflagging energy, and high reputation proved both invaluable and essential. In a real sense, then, this book is another progeny of Harold Wood's endeavors.

Although work began in 1979, the research reported here largely took place after 1981, under a National Science Foundation grant (SES-8024565). Additional financial support, and much appreciated recognition, was provided later by an Ohio State University Distinguished Scholar Award (1984) and a Guggenheim Fellowship (1986). Also important was the time provided by an Ohio State University Professional Leave. Finally, I owe a great deal to John N. Rayner, Chair of Geography, and Joan Huber, Dean of Social and Behavioral Sciences at Ohio State University who have fostered a stimulating and supportive academic climate.

This study builds on a broad data base of high quality, the compilation of which would have been impossible without help from many Latin American professionals. In particular, I want to acknowledge Arthur Conning and Abel Packer of Centro Latinoamericano de Demografia in Santiago, Chile; Roy Ryder, formerly with Centro Panamericano de Estudios e Investigaciones Geograficas of Quito, Ecuador, now at the University of Florida Department of Geography; Antonio Ybarra, formerly with the Instituto

Interamericano de Ciencias Agricolas in San Jose, Costa Rica; and in Caracas, Venezuela, Lourdes Rivero, formerly of the government's Cartografia Nacional, Emilio Osorio, formerly with the Universidad Central, and Isbelia Segnini of the Universidad Central and Director of its Instituto de Geografia y Desarrollo Regional.

The initial theme of this research was development influences on migration, a continuation of ideas put forth in my earlier book on innovation diffusion (Brown 1981). But the work went far beyond that, ultimately addressing the nature of development itself. Many persons contributed. Richard Wilkie of the University of Massachusetts suggested important revisions to the initial research design. At Ohio State, graduate research associates included E. Helen Berry, now at Utah State University; Jorge A. Brea, now at Central Michigan University; Kim V.L. England, now at Ohio's Miami University; Andrew R. Goetz, now at the University of Denver; John Paul Jones III, now at the University of Kentucky; Victoria A. Lawson, now at the University of Washington; Rita Schneider, now at the Free University of Berlin; and Ph.D. candidate Daniel F. Wagner. Faculty research associates, all of whom were affiliated with Ohio State at one time, include Evangelos Falaris, University of Delaware; Janet E. Kodras, Florida State University; Franz-Michael Rundquist, Lund University; W. Randy Smith, Ohio State University; and Frank C. Stetzer, University of Wisconsin-Milwaukee. Others making important contributions to this book are Kevin R. Cox, Nancy Ettlinger, Thomas Klak, Linda Lobao, and Rodrigo Sierra of Ohio State University, and anonymous reviewers.

Diagrams and maps were prepared by Dolores DeMers and Marilyn Raphael; tables by Aly Bradley, Rhoda Rychlink, and Monica Short. Technical assistance in manuscript production was provided by Jay Sandhu, and production itself utilized facilities of Ohio State University's Geographic Information Systems Laboratory under the direction of Duane Marble.

The efforts of those listed above have been essential to this book, and many other outcomes of the research underlying it. In the process, my own vision and understanding grew considerably. I hope my co-workers experienced personal and professional benefits of similar magnitude.

This book is dedicated to Allan P. Jaffe, who died in March 1987. That Jaf was a friend since college is reason enough for the gesture. But he also was a person who helped many to grow and gain self-respect, a person who practiced development at the personal level. As coordinator of Preservation Hall and a musician in his own right, Jaffe expanded the audience for traditional New Orleans jazz, but with care towards not compromising its integrity or that of the performers. With music as a medium, he also instilled pride in the community; for example, by providing jazz instruments to a French Quarter primary school and to individual youth, by encouraging teenagers to form brass marching bands, thus revitalizing a New Orleans tradition, by fostering traditional jazz among young adult musicians, and by launching a school for New Orleans music. With respect to individual musicians and their families, Jaffe's efforts and readiness to help extended far beyond the expected.

In the words of 87-year-old clarinetist Willie Humphrey, "He was a good man", and though Allan would strongly deny it, a development expert in the most basic sense.

L.A. Brown
Columbus, August 1989

1   Introduction

Since World War II, but with roots preceding that time, development has been a major world concern (Arndt 1987). Initial efforts were directed to Europe's reconstruction through the Marshall Plan, European Economic Community, and related arrangements. Attention then turned to the Third World, giving rise to the World Bank; regionally-focussed institutions such as the United Nations Economic Commission for Latin America and the Latin American Free Trade Association; and national organizations of which the United States Agency for International Development and Canada's International Development Agency are examples. Early development efforts were defined around largely economic goals that included increased income and productive capacity, economic independence, and as the engine of change, industrialization and/or trade. Later, goals were broadened to more strongly emphasize equity considerations, distributional aspects of economic growth, basic human needs, and conditions for self-fulfillment.

Despite these good intentions, places continue to differ considerably in characteristics reflecting, or affected by, development – so that Third World landscapes exhibit considerable *unevenness* when development indices are mapped. This regularity holds at all spatial scales; whether the comparison involves small areas, regions within a nation (e.g., Knight and Newman 1976: ch. 6), nations themselves (e.g., Ginsburg 1961), or national groupings (e.g., Jackson and Hudman 1986: ch. 1). Hence, development provides a fertile ground for geographic inquiry, particularly in light of our long and considerable experience in studying areal differentiation, the evolution of place characteristics, spatial interaction, place linkages through which change is transmitted, and the role of such ingredients in societal processes.

Closely intertwined with development is migration, another inherently geographic topic. Population movements within the Third World reflect linkages between areas and place differences in characteristics such as employment opportunities, wage levels, and amenities. The considerable attention given to this phenomena partly reflects its consequences. One view is that migration is detrimental because rural-to-urban flows are chronically large, contribute substantially to high rates of urbanization, and factor into a number of social and economic problems. Others emphasize positive aspects; for example, that migration enhances the efficiency of Third World labor markets, acts as a safety valve to relieve population pressure, and has been an important element in the evolution towards contemporary economies.

The role of development in Third World migration processes was the author's initial concern. Portions of the book which derive from this focus are briefly outlined below under the heading Development Effects on Population Movements in Third World Settings. But the inquiry ultimately broadened to include the fundamental question, 'What is development?', such that Rethinking Third World Development became a theme in its own right. These two topics encompass many of the study's substantive contributions. Other aspects are highlighted in this chapter's final section on the Research Odyssey portrayed in this book.

Development effects on population movements in Third World settings

Since the mid-1960s, migration within Third World nations has been viewed primarily as a response to job opportunity and wage incentives, and to communication or information linkages between places through which incentive differentials become known. Dissatisfaction with that conceptual framework began to surface in the late 1970s and early 1980s. As a proponent of revision during this period, the author argued that 'conventional' models represent an uncritical application of a First World framework to Third World settings; that the relevance of model components is affected by local variations in development (or socioeconomic structure); and therefore, that *place characteristics related to development* should be an *integral ingredient of migration studies*.

Chapter Three presents this argument and associated research. A 'development paradigm of migration' is articulated, followed by empirical analyses of *aggregate* flows within Costa Rica. These show place-to-place variation in the role of migration factors, and attribute this to local occurrences that are elements of development, or of socioeconomic change.

Development differentials between places also affect the behavior of *individuals*. This is demonstrated by reference to Venezuela. **Chapter Four** focusses on migration; while **Chapter Five** focusses on the *labor market experiences* of educational attainment, labor force participation, and wages received. These chapters show that place characteristics associated with development have both direct effects on individual behavior and indirect effects by altering the role of personal attributes. To illustrate, educational opportunity increased with Venezuela's overall development, rendering a direct effect on individual educational attainment (and cognates such as skill level and achievement motivation). Employment opportunity also increased, but in a more restricted, less widespread, spatial pattern. This exacerbated locational disparities between employers and the labor force so that individual migration became more likely, another direct effect of development. Further, because these locational disparities were especially marked for women and the better educated, the personal attributes of gender and educational attainment had a stronger role in migration from less developed settings, an indirect effect of development.

Chapters Three through Five establish the position that place characteristics associated with development play a major role in migration and cognate processes, both in the aggregate and among individuals. These are, however, interconnected phenomena – individual behaviors lead to aggregate change which, when considered for a range of societal processes, alters place characteristics, in turn affecting individual behavior, and so on. **Chapters Six and Seven** highlight this interconnectedness. They focus on *regional change* in Ecuador, and take an approach that embodies the book's view of Third World development, to which attention now turns.

Rethinking Third World development

Development has been a topic of scholarly writing for more than three decades, including an extensive critique related to the introduction of political economy perspectives.[1] Accordingly, the author anticipated a straightforward task in this aspect of the research. It became increasingly evident, however, that established (or conventional) conceptualizations of development were not suitable for understanding the role of place in migration processes at *local* levels, where the author's efforts were focussed. As a result, the question 'what is development?', initially a background exercise, emerged as a major topic.

The need to rethink development stems from a number of

issues. Conventional conceptualizations are concerned with aggregates such as the First and Third World, or broad regional differences within nations such as core—periphery. This is overly general given an interest in understanding Third World locales at *ground level*. Paradigmatic portrayals of development also are at variance with the author's knowledge of Third World settings, gained in part through being there. They omit many elements that are integral to local reality and, more critically, do not provide a framework within which ground-level detail can be accumulated. Finally, and related to the preceding points, the author has become impatient with paradigmatic abstractions that inform on broad tendencies and forces, but provide few particulars and restrict our perception. While highlighting the similarity between places has been important, a more current need centers on place differences and how they may be incorporated into a broad understanding of development or Third World change. Generalization remains the objective, but generalization based on familiarity with, and recognition of, what goes on in the area(s) being studied.

Formulating a general framework that satisfied the author's sense of Third World development, is grounded in and complements earlier literature, and could be an appropriate guide for future research proved a formidable task. A means of resolving contradictions between established views of development and the author's perception of Third World reality grew out of his research on migration and labor market experiences, summarized above. But the framework itself became evident only in the course of writing this book, when findings of the various research modules were set beside one another. Refinement is still ongoing.

The current version is presented in **Chapter Two** which advocates that development aspects of Third World landscapes and their change over time represent the *local articulation of world economic and political conditions, donor-nation actions, and policies of Third World governments* themselves. Spatial variation in the manifestation of these forces occurs through their interaction with local conditions. Hence, to understand development, or regional change, an appropriate analytical framework considers the aforementioned external, or exogenous, forces; the way they are *mediated by local conditions*; net effects on enterprises and individuals (and/or their response); and how the conjunction of these elements translates into socioeconomic change at local, regional, and national scales.

Seeing exogenous forces as critical elements of Third World

development is not new in itself, but because many view these as secondary to other dynamics, the emphasis is significant. Stressing the *local articulation* of exogenous forces represents an even greater departure from the usual approach. To illustrate, it is generally agreed that macroeconomic conditions and national policies, even those oriented towards rural sectors, ultimately benefit (large) urban areas (e.g., Lentnek 1980; Lipton 1976; Todaro and Stilkind 1981). But 'urban bias' is primarily depicted as a general tendency or in terms of gross distinctions such as rural–urban; not in terms of its manifestation in, or impact on, particular locales. Similarly, research associated with neoclassical and political economy conceptualizations tends to focus on paradigm-related issues of a general nature and, in spatial matters, broad regional effects rather than local ones.

Chapter Two addresses issues such as the above, elaborates the approach to development advocated by this book, and provides a synopsis of conventional views of development, i.e., neoclassical and political economy conceptualizations. Chapters Three, Four, and Five accumulate evidence that local occurrences, or development events, entering into migration and labor market experiences invariably reflect world economic and political conditions, donor-nation actions, and government policies. But the primary objective of Chapters Three through Five is to examine population movements in Third World settings; contributions to the development argument are a secondary gain.

By contrast, with a foundation firmly in place, Chapters Six and Seven focus explicitly on external forces representing the development framework advocated by this book, the means of their articulation in local areas, and their role in population movements and regional change in Ecuador. Policies considered include land reform, agricultural pricing, agricultural credit, monetary exchange, and other elements of Ecuador's import substitution industrialization program. The time frame includes economic expansion in the early 1970s and contraction by the early 1980s, both related to world petroleum markets. Chapter Six examines *circulation* and *migration* among *individuals* from the rural Sierra; Chapter Seven examines *socioeconomic change* among *local areas* comprising all of (mainland) Ecuador. Both chapters demonstrate that the effects of national policies and world conditions on individuals and locales are highly variable; and that an important determinant of effect is local socioeconomic structures or place characteristics related to development, which function as mediating agents.

The *argument* thus comes *full circle*.   Early chapters establish that place characteristics associated with development play an important role in migration and labor market experiences; and that these characteristics represent the aforementioned external forces. But direct examination of selected forces, in Chapters Six and Seven, indicates their translation into regional change is mediated by local socioeconomic structures, which in turn reflect a variety of societal processes such as migration.   There is, then, an *interdependence* in that place characteristics related to development affect societal processes, which in turn affect place characteristics, and so on.

Having closed the circle of inquiry, **Chapter Eight** considers *research implications* of the approach to development being advocated.   'What have we learned?' and 'Where do we go from here?' are its central concerns.

As a final note to the development theme, this book's empirical studies give greater attention to national policy than to world conditions or donor-nation actions.   This is partly accidental, partly a reflection of data availability, and partly because policy represents actions by Third World nations themselves rather than an external entity.   But the analytical approach, reasoning, and conclusions related to policy pertain equally well to donor-nation actions and world economic and political conditions.

The odyssey of research and related features of the study

Research often takes one far afield from the original question, creating a journey of unexpected dimensions.   In the present instance, a central aspect of that question was 'development'. Because this has received an enormous amount of attention, the author naively anticipated little need for further inquiry; to simply take concepts, methodologies, indices, and related accoutrements 'off the shelf' seemed a likely option.   Instead, as described in the preceding section, articulating and implementing a satisfactory development framework emerged as the most challenging issue. This partially offset the original research focus, development effects on migration and related population processes, but that focus also provided the medium by which a more satisfactory understanding of development was achieved.

In telling the story, the order of inquiry has been maintained except for those portions of Chapter Two which set out the author's current view of development. One consideration underlying this arrangement of the material is that portraying the research odyssey by which ideas emerge has intrinsic value. In particular, presenting the argument in this way elucidates the role of *discrete*, *self-contained studies* in building to a broad theme. Each study also provides detailed literature reviews, suitable research designs, methodological suggestions, findings, and a platform for future, state-of-the-art research on specific topics. Hence, these research modules are both elements of a broad argument and an *important feature* of the book in their own right.

That this research odyssey *interweaves*, or *blends*, broad concerns is another important feature. Population movements and labor market experiences cannot be properly understood without reference to development, but they also provide a means for better understanding development. Place also plays a role in that both sides of the equation are affected by local conditions, which introduces the contingency aspect of societal processes.

Finally, this book *bridges perspectives*, taking them as complementary rather than competing. Relevant in this regard is the author's odyssey as a professional. His training emphasized empirical analysis, modeling, paradigmatic thinking, and generalization; i.e., nomothetic, positivistic social science. But in subsequent work, including that reported here, he has sought out a middle ground wherein nomothetic and idiographic perspectives are integrated with the objective of arriving at a substantive, often particularistic, understanding, while nevertheless aiming for generalization. Other persons steeped in nomothetic traditions, including political economy ones, might feel this inquiry moves too far from, misinterprets, or underestimates the breadth and adaptability of earlier perspectives. Alternatively, a person from traditions that emphasize case study detail rather than generalization, such as historical or regional geography, might feel the inquiry does not move far enough towards the idiographic. Similar observations apply to persons from social science disciplines other than geography. Yet, the tension created by differing perspectives is essential to extending knowledge.

Many aspects of this book illustrate the gain from bridging perspectives; for example, its conceptualization of development and analyses of migration processes. Another, less obvious gain lies in the use of *census data*. The objective is to draw substantively informed conclusions from statistical analyses of readily available data with broad geographic coverage. Census data provides a basis for generalization because it depicts numerous locales, but many criticize its use because, from their perspective, census variables are not inherently capable of providing an understanding of socioeconomic processes. Alternatives include surveys or case studies designed around an understanding of process, but these usually represent only selected (or a single) locales. This book demonstrates throughout that both qualities, understanding and representativeness, can be realized by using qualitative knowledge of places to interpret statistical findings from census data.

In this regard, *place knowledge* becomes an essential ingredient of research. Returning to the geography of place is not an end in itself, but a means for understanding societal processes, human behavior, and the role of place therein. By focussing on the interaction between external forces and local conditions, for example, we see that place characteristics associated with development have effects in their own right, often greater than variables conventionally considered, and also affect the role, or relative importance, of other variables. Hence, the reader should come away with an enhanced sensitivity to the significance of place in human behavior.

To close with a condensed summary, this book encompasses the following. Major topical concerns are the impact of development on population movements in Third World settings and, a related issue, the nature of development itself. As an offshoot of these concerns, labor market experiences and regional change also are considered. Empirical examination of these topics centers on Latin America – in particular Costa Rica, Venezuela, and Ecuador, but research from other Third World areas is drawn on extensively. Concerning spatial scale, data analyses focus on small areal units or individuals, but as elements of national (in one case regional) aggregates; verbal discourse employs examples representing all spatial scales. Highlighted, however, are atypical localized occurrences, differences between local areal units, and heterogeneity within regions. Discrete, self-contained studies are the primary means of addressing each topic. In addition to providing substantive understanding and platforms for future research, these studies contribute to the book's methodological dimension. One aspect of this is demonstrating the application of several research designs and related statistical procedures. That *qualitative* knowledge of places is critical to understanding statistical findings also is shown. In this regard, we are faced with a continuum at one end of which is immersion in the details of a particular locale to the exclusion of identifying its generalizable aspects, and at the other end, interpreting data without incorporating an understanding of the place(s) they represent. This book contends the balance between these extremes should shift so that place knowledge becomes a more integral element of research methodologies.

2　What is Third World development?

Discourses on Asia, Africa, and Latin America are replete with terms such as Third World development, economic development, uneven development, regional development, and the like. Because these terms are commonly used without elaboration, one might assume that 'development' is well defined, its meaning agreed upon. But in fact, two people referring to Third World development often mean quite different things and hold little inclination towards compromising their position.

Nevertheless, there is growing consensus that the numerous views of development reduce to a few basic forms. In this spirit, Wilber and Jameson (1988) propose a two-fold classification of development frameworks as either orthodox or political economy. They also state a need for development research to move beyond established paradigms, saying (1988: 22, emphasis added) that the "concrete record...challenges us to **rethink** our approach to development".

This chapter follows a parallel path. It first surveys established, or conventional, perspectives on Third World development, in a section titled A Synopsis of Development Paradigms. This provides background material for empirical studies in subsequent chapters as well as a basis for elaborating the author's perspective – Third World Development as the Local Articulation of World Economic and Political Conditions, Donor-Nation Actions, and Government Policies. That framework draws on and amalgamates well-known elements, but differs from established paradigms in orientation, research approach, and factors emphasized. Hence, the framework conforms with Wilber and Jameson's exhortation towards 'rethinking' and is, in this sense, a *timely* statement. Finally, the section Development or Modernization? draws a comparison between socioeconomic change experienced earlier by developed nations, such as the United States, and that ongoing in today's Third World. This comparison, and the author's perspective, support a shift in research strategies for studying development which is mirrored in subsequent chapters.

A synopsis of development paradigms

Social science research has put forth a number of frameworks that identify conditions related to development and describe the process overall.   These primarily focus on characteristics of nations or regions within nations, but give some attention to personal attributes.   Four reasonably distinct approaches may be identified: Stage Formulations, Dual Society Models, Human Resource Approaches, and Political Economy Perspectives. Attention also is given to the Latin American School of Development which provides a bridge to the framework put forth by this book.

Stage formulations

An early representation of the development process is Rostow's (1960) stages of growth model.   Drawing directly on economic history, five sequential stages characterize national development: traditional society, preconditions for take-off into self-sustaining growth, take-off, drive to maturity, and high mass consumption (Figure 2.1.A).   The engine of take-off is manufacturing, or secondary activity.   Preconditions for take-off include an extensive public infrastructure and an entrepreneurial mentality within society.   Savings and capital accumulation also are critical, a theme emphasized in the Harrod-Domar model (Todaro 1985: 63–7).   Progress of various nations along Rostow's continuum up to 1960 (Figure 2.1.B) provides a spatial frame of reference for the stage approach and an illustration of its application.

Although stage thinking is often seen as dated, many of its elements are found in subsequent models: for example, labor as a resource on which incipient industry can build; emergence of an entrepreneurial, or achievement-motivated mentality; and technology and innovation as a growth stimulus (or deterrent).   Further, infrastructure proliferation, capital accumulation, and industrial buildup remain at the heart of today's development policies. Finally, the ebullient but misplaced optimism of stage models should be noted.   Rostow's formulation, for example, is billed as a non-communist manifesto, but it naively puts aside internal structure and world system aspects that restrict nations in economic decision-making, limit the range of potential change, and retard development.

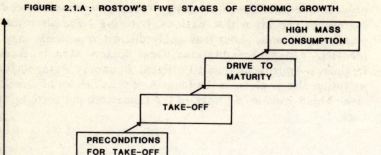

FIGURE 2.1.A : ROSTOW'S FIVE STAGES OF ECONOMIC GROWTH

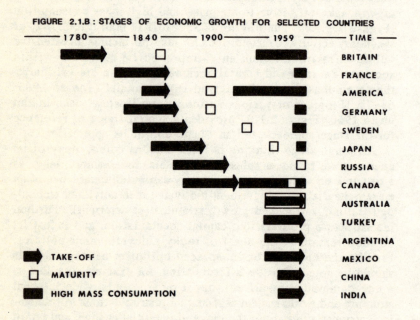

FIGURE 2.1.B : STAGES OF ECONOMIC GROWTH FOR SELECTED COUNTRIES

Figure 2.1 Development stages as seen by Rostow (1960)

Dual society models

The major concern of dual society models is the balance between contemporary and traditional elements in national and international economies, and processes by which it shifts towards modernity. These formulations are articulated primarily in terms of the internal structure of a single nation.

In economics, for example, the *two-sector growth model* (Fei and Ranis 1964; Lewis 1954; Todaro 1976: 21–5; 1985: 67–72) posits a traditional, rural subsistence sector and a modern, highly productive urban sector. The mechanisms of structural transformation are industrial expansion and surplus rural labor that can be withdrawn without loss of output (i.e., it has zero or near zero marginal productivity). Specifically, industry expands by importing labor from the rural sector at a fixed, low wage (since it is surplus); reinvests profits thereon to further expand the urban industrial base; which again is fueled by imported labor, etc. This process terminates when labor surplus is exhausted and wages rise in response to supply/demand conditions. Meanwhile, industrial expansion (based on the competitive advantage provided by low wages) provides a catalyst for transforming society from traditional to modern. Two-sector growth formulations thus provide a rationale for industrial buildup and labor exploitation; establish rural–urban migration as an integral mechanism of development; and view urban areas as the source of economic growth.

Sociology's dual society model is *modernization theory* (Rogers 1969). Traditional and modern are defined in terms of personal attitudes, which change through media communications, interpersonal information flows, demonstration effects, and education. Particular emphasis is given to bringing about achievement motivation (McClelland 1961) and/or an entrepreneurial mentality (Hagen 1962).

Dual society models with an explicitly *spatial* frame of reference include *core–periphery* (Friedmann 1966, 1972, 1973, 1975) and *growth center* (Hansen 1971: ch. 2; 1972; Richardson 1976; Richardson and Richardson 1975) formulations; applicable, respectively, to national territories and to a city and its hinterland. Elaborated by regional planners and geographers, these divide geographic space into a dynamic, propulsive, modern core (or growth center) distinguished by secondary and quaternary activity, and a traditional periphery (or hinterland) dominated by primary activity. These areas are linked through two mechanisms. Backwash or polarization effects drain the periphery through migration, capital movements, trade flows (which occur at terms disfavorable to the periphery), and the like. Simultaneously, there also are spread or trickle-down effects which build up the hinterland through income remittances, innovation diffusion, infrastructure proliferation, education, establishment of secondary or service activities, etc. Polarization effects initially outweigh trickle-down, leading to core domination. This disparity increases in a 'circular and cumulative' fashion (Hirschman 1958; Myrdal 1957) due to the importance of spatially concentrated agglomeration economies. Eventually, however, trickle-down forces gain strength, and the periphery's rate of growth exceeds the core's, a phenomenon termed 'polarization reversal' (Richardson 1980; Townroe and Keen 1984).

When in this cycle does a particular locale benefit from polarization reversal? Relevant elements include its distance from the core and other urban agglomerations; size and growth rate of the core; characteristics of urban, transport, and communication systems in the periphery; and the spatial distribution of sociopolitical power (Gaile 1980). Hence, the periphery is not uniform, but may be distinguished into downward and upward transitional locales (Figure 2.2). The former include traditional, perhaps once prosperous, areas that are presently stagnated. An area may be upward transitional because it is located near the core or in a development corridor (between two growing locales), because it has natural resource endowments, or because it is a frontier settlement with economic potential or strategic significance (Berry, Conkling, and Ray 1987: ch. 16; Stohr 1975).

Some general characteristics of dual society models should be noted. First, they divide the national territory into two segments; one modern, the other traditional. Second, development consists of modern sector expansion, and this is linked with traditional sector erosion. Third, diffusion through society or across the geographic landscape is a basic development process; accordingly, diffusion has received a good deal of attention by both academics and policy-makers (Brown 1981: ch. 8). Fourth, dual society models advocate a centrally-focussed, trickle-down, or top-down development strategy rather than a grassroots, percolation, or bottom-up approach. Fifth, development is accompanied by changes in economic, social, and spatial structures. The international comparisons of Chenery and Syrquin (1975), for example,

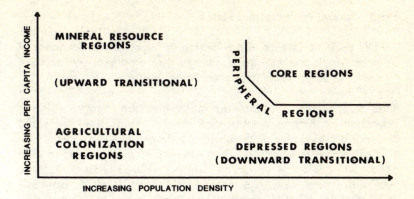

Figure 2.2 Core and periphery distinctions

show employment shifts out of primary activity and into manufacturing and services; an increase in capital accumulation as indicated by outlays for schooling, investment, savings, and human resources as a percent of gross domestic product; and consumption pattern changes wherein the percent of gross domestic product allocated to non-food consumption increases and the percent allocated to food consumption decreases (also see Chenery 1979). Concerning spatial structure, core–periphery or growth pole–hinterland patterns have been found for development in Ghana (McNulty 1969, 1976), Kenya (Soja 1968), Sierra Leone (Riddell 1970, 1976), Tanzania (Gould 1976), Chile (Berry 1969; Pedersen 1975), and Mexico (Scott 1982). In general terms, development is greatest at or near major urban centers, declines monotonically with distance from those locales, is lowest in remote rural areas, and changes over time in a spatial diffusion manner.

Human resource approaches

Whereas stage and dual society models emphasize structural conditions, the human resource perspective focusses on individuals and their ability to effect development (largely a bottom-up rather than trickle-down approach). This holds that growth results from improved labor productivity, personal skills, motivation to achieve, and ability to exploit opportunities, leading to locally spawned, spontaneous economic activities. Schultz's (1980: 640) Nobel

Prize address, for example, states

> The decisive factors of production in improving the welfare of
> poor people are not space, energy, and cropland; the decisive
> factor is improvement in population quality.

This perspective complements modernization theory, which is
concerned with transforming traditional to modern attitudes and
creating achievement motivation or entrepreneurial values.

Education is considered the major mechanism by which
population quality is improved, and accordingly, its effects have
received a good deal of attention (Blaug 1973, 1976; Bowman
1980; Colclough 1982; Layard and Psacharopoulos 1974; Lockheed, Jamison, and Lau 1980; Psacharopoulos 1980; Psacharopoulos and Hinchliffe 1973; Psacharopoulos and Woodhall 1985;
Todaro 1985: ch. 11; Yotopoulos and Nugent 1976: ch. 11).
Labor force participation, small-scale enterprise, and attendant
apprentice systems also are important (Anderson 1982; Anderson
and Leiserson 1980; Blaug 1973, 1976; Cortes, Berry, and Ishaq
1987; Gordon 1978; Ho 1980; Layard and Psacharopoulos 1974;
Little, Mazumdar, and Page 1987; Page 1979; Page and Steel
1984; Schneider-Sliwa and Brown 1986; Squire 1979; Standing
1982). The spatial variation of these elements has received
attention, but rarely in a human resources frame of reference.
An exception is Brown and Kodras (1987) and Brown and Lawson
(1989) who show that core areas in Venezuela supply quality
human resources to the periphery and that polarization reversal
may be viewed in human resource terms.

Basic needs (Ghosh 1984; Lisk 1977) is another aspect of the
human resource approach to development. Education improves
health, nutrition, household living conditions, and other aspects of
basic needs which, aside from bettering individual lives, contribute
to elevating labor productivity (Bowman 1980; Colclough 1982;
Fields 1980; Noor 1981; Todaro 1985, ch. 11). Such improvements derive from value shifts as well as knowledge; however,
higher incomes related to educational attainment also play a role
by enabling purchase of better food, housing, medical care,
sanitation services, and the like. Finally, because education leads
to entrance of females into the labor force and increased social
and geographic mobility, it is associated with lower fertility
(Cochrane 1979; Colclough 1982; Findley 1977; Findley, Gundlach, Kent, and Rhoda 1979), which in turn eases non-productive

demands on national economies. On the basis of these and similar effects, public expenditures for education, health, and nutrition often are advocated as providing high returns to society (World Bank 1980).

Political economy perspectives

Dependency, Marxist, and historical-structural formulations function largely as a critique of stage, dual society, and human resource approaches, and of the development strategies they imply. A basic tenet of political economy perspectives is that development and underdevelopment are structurally interdependent in a manner whereby interests of the politically dominant core are best served by limiting actions of the periphery. At an intranational scale, for example, growth in the core depends on restricting economic activity of the periphery in order to continue receiving less costly labor, capital, and other polarization flows. An international example, recently popularized in the movie *Gandhi*, is Britain's restriction of textile and salt production in India, thriving indigenous industries prior to colonization. This both expanded the market for British-produced goods and operated as a form of political control. O'Brien (1975: 15–16) provides another international example:

> the incorporation of Latin America into the emerging world capitalist economy, first through a direct colonial administration and then, more subtly, through free trade, ensured that Latin American production was geared towards producing exports for the dominant economies, and the political and social system ensured that the gains from this were divided between a small Latin American class (who used their gains for importing luxury consumer goods rather than diversifying investment) and the dominant metropolitan countries... Thus the determinant of the growth and structure of the Latin American socioeconomic formations remained largely exogenous to Latin America... With the collapse of the world capitalist economic order in 1929, Latin America began a process of trying to lessen her external vulnerability on the world markets. This entailed both radical shifts in political alliances and power..., and the use of various policies... This drive (for national development) took the form of import substitution

industrialization. But instead of national development, the result was but another form of dependency.

More generally, political economy perspectives contrast with others in five ways. First, they take a viewpoint wherein economic and social outcomes are conditioned by power relationships. Second, the development of less developed areas is thus limited by their dependent status, and characterized by a clash, rather than harmony, of interests. Third, it further follows that development is not inherently good, not necessarily an end state that individuals, regions, and nations should aspire to. Fourth, under political economy thinking underdevelopment is a consequence of the world system; other perspectives are more inclined to see cause in terms of Third World countries (regions, individuals) themselves. Related to this, political economy approaches call attention to socioeconomic structure and its effects more stridently, more distinctly, than do other perspectives. Finally, political-social-economic relationships are historically determined, in part through idiosyncratic events. Methodologically, therefore, each situation is considered both in terms of general processes and in terms of its unique, historical context.

To exemplify these distinctions, consider the internal spatial structure of Third World countries (Fair 1982: 25–36). Non-political economy (or *orthodox*) approaches explain urban primacy, regional polarization, and dendritic, core-focussed transportation networks in terms of agglomeration economies needed for 'take-off' or modernization; i.e., as integral elements of economic change. Under political economy reasoning these same characteristics reflect the external orientation of production, and control by colonial powers and coopted internal elites. Similarly, transformation of pre-colonial West Africa's urban system (Mabogunje 1968, 1981: chs. 7 and 12) into that of today (Figure 2.3), or the 'ideal–typical' sequence of transportation development (Taaffe, Morrill, and Gould 1963) (Figure 2.4) represents two forces – the diffusion of modernity and economic change, an orthodox explanation; but also, colonial penetration and domination, a political economy explanation.

These examples illustrate the distinction between political economy and other perspectives. But they also demonstrate that explanations associated with each perspectives may be seen as *complementary* rather than competing. This important point is returned to in delineating the book's perspective on development.

FIGURE 2.3.A : WEST AFRICAN TOWNS AND TRADE ROUTES,
SIXTEENTH CENTURY

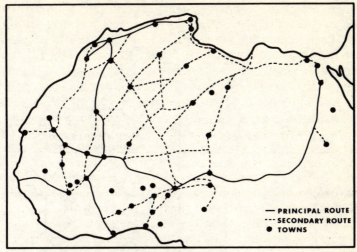

SOURCE: ADAPTED FROM MABOGUNJE (1968: 48)

FIGURE 2.3.B : WEST AFRICAN URBAN CENTERS, 1981

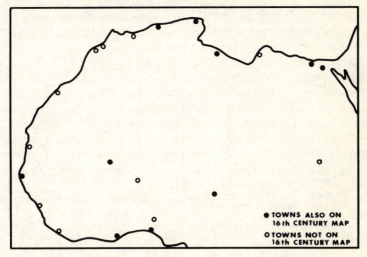

Figure 2.3 Impact of colonization on the urban system of West Africa

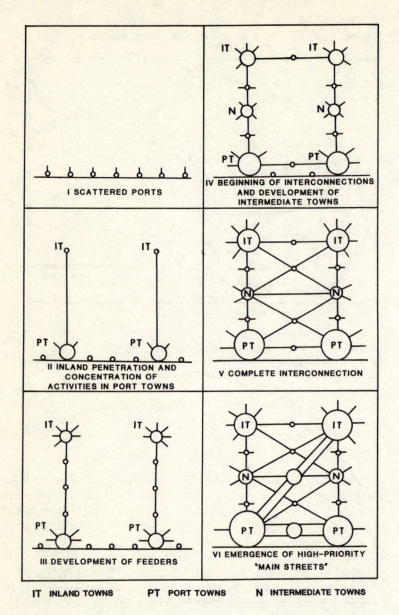

Figure 2.4 Ideal–typical sequence of transportation development in a developing country

The Latin American school of development

Latin American contributions to development thinking emphasize the role of macroeconomic/political conditions and national policies, and often use place-specific examples to elucidate how such forces operate. In this respect, they provide a bridge to the author's perspective on development, which advocates an analogous approach. That empirical studies in Chapters Three through Seven focus on Latin American settings is another reason for highlighting the school.

A comprehensive review and critique of the Latin American school is provided by Kay (1989). In general, it falls under the political economy rubric and represents a reaction to First World or core perspectives such as the stage, dual society, and human resource approaches outlined above. A guiding theme was

> to decolonize the social sciences...to see the periphery through their own eyes and develop alternative theories which provide a more truthful interpretation of their reality. (Kay 1989: 14)

In its early years, a major impetus was the 'structuralist' perspective of social scientists affiliated with the United Nation's Economic Commission of Latin America, under the direction of Raul Prebisch. Kay (1989: 25) observes that

> ECLA's initial concern lay in discovering the major obstacles to development in Latin America and suggesting policies to overcome these. Thus, the concern was more practical than scholastic.

Import substitution industrialization strategies dominated structuralist thinking at the start, motivated in part by the need to counteract practices associated with then prevailing orthodox and neoclassical views. While the prescription would shift, policy concerns remained at center stage.

Other perspectives of the Latin American school include Internal Colonialism, Marginality, and Dependency (Kay 1989). Central to these lines of inquiry, as well as Structuralism, is the role of macroeconomic/political conditions, power relationships, and Latin America's position in the world order. To establish the relevance of such elements and illustrate how they worked, the Latin American school (like other political economy efforts) often relied on place-specific, historical examples. Hence, Dos Santos (1974) examines Brazil from colonization through its 'economic miracle' of the early 1970s; Frank (1967) employs case studies from the economic and social histories of Chile and Brazil; Furtado (1976) and Cardoso and Faletto (1979) examine Latin American development overall but draw on detailed examples from various historical eras of its major countries (e.g., Argentina, Bolivia, Brazil, Chile, Colombia, Mexico, Peru, Uruguay, Venezuela).

Although these studies are multifaceted and empirically rich, it is their political economy dimension that has most impacted development thinking and constitutes a distinguishing trait. This may be traced to enthusiasm over a newly emerging paradigm. But also, while some writers of the Latin American school simply note that obstacles to development tend to be institutional (e.g., Furtado 1976), the majority overtly display a paradigmatic orientation, often in strident terms. For example, Dos Santos (1974: 418, emphasis added) states

> an analysis of Brazil's historical evolution becomes the elaboration of a **correct** theory of underdevelopment and dependency as well as a theory of the social revolution which marks the current history of the Third World.

and (1974: 486)

> The struggle between socialism, as the only popular solution, and fascism, as the only capitalist alternative, will continue to be the key to the Brazilian historical process.

Similarly, Cardoso and Faletto (1979: 172) say

> theoretical schemes concerning the formation of capitalist society in present-day developed countries are of little use in understanding the situation in Latin American countries. Not only the historical moment but also the structural conditions of development and society are different.

and Frank (1986: 122) echoes with

> They will not be able to accomplish these goals by importing sterile stereotypes from the metropolis which do not correspond

to their satellite economic reality and do not respond to their
liberating political needs.

More generally, Kahl (1976: 196) notes in his study of Germani,
Casanova, and Cardoso

> scholars considered in this book...define crucial problems...in
> terms of the most important practical issues their societies face.
> For them, development is not just a question of intellectual
> disputation, but also a matter of national survival... However,
> the very selection of topics for study has an ideological and
> theoretical bias. Although these scholars are more concerned
> with an exploration of national reality than they are in testing
> academic theories of development, they nevertheless cannot
> escape certain theoretical choices that guide their research.

From today's vantage point, political economy emphases within
the Latin American school may be downplayed as a product of
the academic dialectic of the time, a point discussed later in the
chapter. But this emphasis has diverted attention from other
issues. Especially critical is the need to interpret empirical
evidence in its own right and modify (or derive) theory accordingly, rather than operating within an established conceptualization.
Hence, the author feels strongly that those elements of the Latin
American school which usually have been subordinated to theoretical concerns should instead be at the center of research designs;
e.g., macroeconomic/political conditions, national policies, and the
importance of place knowledge in obtaining an understanding of
development.
    Similar sentiments have been expressed by others. In address-
ing the present 'impasse' in development studies, Booth (1985:
775) notes that data interpretation often has been influenced by an
investigator's commitment to particular political economy concepts, leading to

> persistence in analyzing development problems in certain kinds
> of ways even when they can be explained well or better in
> other terms.

And Kay (1989: 211) notes that

> structuralist and dependency analysts have to undertake more

studies of the smaller or micro units of a country...and to consider the possibility and feasibility of a variety of styles and paths of development. It is only at a very high level of abstraction and simplification that dichotomies such as capitalism and socialism are valid.

An alternative perspective: Third World development as the local articulation of world economic and political conditions, donor-nation actions, and government policies[1]

Research on Third World development may be categorized by the spatial scale to which it applies. At a highly aggregate level are studies concerned with relationships between the Developed and Developing World. World system, dependency, or international division of labor perspectives exemplify this thrust. Representing a middle level of aggregation are cross-national studies that focus on sectoral shifts in the labor force related to economic maturation, stages-of-growth constructs, or development differentials among Third World countries. Next on the continuum is research concerned with regional differentiation within a Third World nation; associated with this are core—periphery, two-sector growth, modernization theory, human capital, and internal dependency models.

Among conceptualizations just cited, and reviewed earlier, some highlight conditions for self-sustaining, autonomous growth or mechanisms for bringing it about; others stress factors inhibiting such growth. Also, forces ascribed to greater levels of aggregation have effects at lesser levels. And when taken as complementary perspectives, conventional frameworks add immeasurably to our understanding of development as a broad-gauged process affecting a range of spatial scales.

But a serious *question* arises if focus shifts from development itself to understanding socioeconomic landscapes of *particular* Third World settings and changes therein; i.e., to ground-level reality at sub-national spatial scales. Then, study area details become more central to social science procedures, leading to a realization that the essentials of change (or stasis) are inadequately addressed by conventional development frameworks, even when their complementary points of view are melded together.[2]

In response, this book advocates a research protocol based on the contention that development aspects of Third World landscapes and their change over time represent the local articulation of world economic and political conditions, actions of donor nations, and policies of Third World governments themselves. As a preamble to elaborating this perspective, note the following.

First, emphasis is on *change*, not development. The issue is that Third World landscapes both advance and decline, whereas development implies only the former.

Second, change represents an *interplay* between forces external (or exogenous) to a locale and characteristics of the locale itself. The author's predilection concerns how an exogenous force may have substantially different impacts among places, depending on local circumstances/responses which act as a conditioning agent, and the resultant (spatial) variation in outcomes. Emphasis on other aspects of this nexus may be preferred by others.

Third, *place knowledge* is deemed critical to the task of understanding change. This entails a substantive understanding of both endogenous characteristics and exogenous forces.

Fourth, conventional development frameworks are seen to account for elements of Third World change, but not the essentials. They also are considered *complementary* to one another. Hence, conventional frameworks are put into perspective, but *not* dismissed.

Fifth, among *exogenous forces*, world economic and political conditions encompass the operation of world commodity markets, multinational corporations, international trade, financial institutions, international loan practices, technology transfers, and the like. Donor-nation actions encompass unilateral assistance related to infrastructure creation, development projects, food and health service provision, education, and job skill acquisition; providers include USAID and the World Bank. Policies of Third World nations encompass import substitution and export-led industrialization measures, social service provision, labor laws, fiscal and taxation practices, and more.

Sixth, the aforementioned categories of exogenous forces are individually identifiable but not independent of one another; they encompass an enormous, perhaps exhaustive, range; and most importantly, the *direction* of research is clearly charted. From a schematic point of view, furthermore, world conditions and donor-nation actions represent relationships between the Developed and Developing World, while national policies represent actions of

Third World nations themselves. Finally, although the importance of exogenous forces is widely accepted, our knowledge of their operation in, and impact on, local areas is minimal.

Seventh, the argument builds on, integrates, and extends several issues recognized as important by other researchers. For example, in addition to earlier observations concerning the Latin American school of development, that the dynamics and source of economic growth are external to Third World economies is discussed by Caporaso and Zare (1981); the interplay between exogenous and endogenous forces is discussed by Browett (1985), Stohr (1975), and Stohr and Taylor (1981); local articulation issues by Lipton (1976) and Massey (1984); the complementarity of established frameworks by Ettema (1979) and Simon (1984); and the reemergence of place as a geographic concern is discussed by Pudup (1988).

But these discourses either embrace a singular (or restricted) epistemological view; have not been explicated in terms of Third World development at the local scale; and/or have gone unheeded in geography and regional science. By contrast, the perspective put forth here is synthetic in approach, moves beyond the confines of established development frameworks, sees exogenous forces as the primary dynamic (rather than secondary to others), and urges research protocols that consider the totality of a locale's experience relative to particular forces of Third World change.

A rationale for the author's point of view is presented in four steps. First, to provide a basis for the argument, we momentarily return to conventional perspectives in the sub-section Paradigmatic Statements Concerning Development: A Precis. Next, a link between Paradigmatic Frameworks and Third World Policies is established. Of the two, policies and related forces better elucidate the reality of Third World settings. To embellish the reality theme, attention turns to Third World Landscapes at Ground Level, which includes commentary on socioeconomic change in developing societies and social science approaches to understanding it. The final segment of the argument sets out our need to Move Beyond Paradigmatic Thinking.

Paradigmatic statements concerning development: a precis

The many conceptualizations, theories, and models of development reduce to a few basic forms. The early portions of this chapter, for example, distinguish Development as Evolution, Dual Society Models, Human Resource Approaches (including Basic Needs), and Political Economy Perspectives (including the Latin American School). Even more simply, Wilber and Jameson (1988) classify development paradigms as either Orthodox (Neoclassical) or Political Economy.[3]

However paradigmatic discourses are classified, in focussing on 'development' they imply that nations (regions) ought to pass from stage to stage, each being higher than its predecessor; i.e., development paradigms are "parables of progress" (Wilber and Jameson 1988: 5–7). Whether orthodox or political economy in orientation, they also see "internally balanced, self-sufficient, autonomous" growth as a (the?) major goal (Phillips 1977: 19). With regard to defining development, some paradigms use highly specific, relatively narrow frames of reference; others use ineffable, evanescent, more general terms; occasionally, development is simply undefined, i.e., taken as a given or as a 'black box' phenomenon.

To elaborate, Wilber and Jameson (1988: 7–14) note that orthodox paradigms see development in terms of specifics such as high mass consumption, greater availability of goods and services, and an open economy. Progress ideally occurs through laissez-faire mechanisms, but these must be augmented by policy due to market imperfections, institutional rigidities, cultural dualism (e.g., traditional versus modern values), and the like. Finally, some orthodox paradigms emphasize economic growth as the key to development; others the distribution of growth benefits ('growth-with-equity').

The latter emphasis also is a concern of political economy perspectives (Wilber and Jameson 1988: 14–22). These tend to focus on the inhibition of economic growth and structural factors related thereto, why development has progressed more in some nations (or areas) than others, conflictual relations among socioeconomic entities, who controls development processes, and ultimately, liberation from oppressive and exploitative relationships. In the end, political economy and orthodox paradigms often address similar phenomena, but from a different perspective. For example, political economy examinations of social dualism might focus on forces giving rise to and perpetuating its existence, whereas inhibiting effects on economic growth might be of greater concern under orthodox approaches.

Orthodox and political economy paradigms also interrelate with popular, more general, observations on development. For example, in the spirit of political economy thinking, Grunig (1971: 581) states

> The most important of these realities (of Third World settings) are the structural, institutional, and social rigidities that must be broken if meaningful development is to ensue.[4]

From a more orthodox perspective, Heilbroner (1967: 31-2) expresses a similar sentiment after saying

> the revolution of rising expectations...conjures up the image of a peasant in some primitive land, leaning on his crude plow and looking to the horizon, where he sees dimly, but for the first time...and that is what is so revolutionary..., the vision of a better life. From this electrifying vision comes the necessary catalysis to change an old and stagnant way of life. The pace of work quickens. Innovations, formerly feared and resisted, are now eagerly accepted.

And in a synthesis of perspectives, Todaro (1985: 83-91) lists three objectives that embrace core values of development: (1) life sustenance by providing basic needs such as food, shelter, health, and safety; (2) self-esteem, individually and nationally, by raising the level of living through higher incomes, more jobs, better education and improved cultural and humanistic values; (3) freedom from servitude by expanding the range of economic and social choices available to individuals and nations. He then says (1985: 85)

> Development must, therefore, be conceived of as a multidimensional process involving major changes in social structures, popular attitudes, and national institutions, as well as the acceleration of economic growth, the reduction of inequality, and the eradication of absolute poverty. Development, in its essence, must represent the whole gamut of change by which an entire social system, tuned to the diverse basic needs and desires of individuals and social groups within that system, moves away from a condition of life widely perceived as unsatisfactory and toward a situation or condition of life regarded as materially and spiritually better.

In short, 'development' has come to mean a plethora of socio-economic conditions; as Arndt (1987:1) notes, perhaps facetiously,

> to encompass almost all facets of the good society, everyman's road to utopia.

Nevertheless, paradigmatic statements concerning development and their more popular variants render a compelling, albeit amorphous, ideal(s).

Paradigmatic frameworks and Third World policies

Policies are a concern of development frameworks, but one that is secondary or derivative. Under orthodox thinking, government actions primarily are catalysts to offset (and correct) market imperfections, institutional rigidities, and more generally, the slow pace of development. Political economy thinking ascribes similar tasks to policies, but tends to view them in terms of lessening inequality or removing shackles that impede growth, rather than as a catalyst; critical appraisals of policy measures also are common.

Yet, if we turn the mirror to focus on national policies themselves, their conceptual underpinnings are conspicuous. Consider import substitution industrialization (actually a set of policies rather than one). Its emphasis on capital formation coincides with stages of growth, two-sector growth, and Harrod-Domar frameworks; its definition and promulgation represent a pragmatic response to economic dynamics of the world system; and the necessity of such response coincides with political economy thinking (Arndt 1987: 54–60, 72–87; Brookfield 1975: 133–65; Furtado 1976: chs. 10–12; Swift 1978: 56–64). Similarly, policies fostering improvements in educational attainment are a human capital strategy (Todaro 1985: ch. 11); economic decentralization reflects core–periphery/growth center and political economy thinking (Lipton 1976; Richardson 1980; Rondinelli 1983: ch. 1; Rondinelli and Ruddle 1978; Stohr 1981; Todaro and Stilkind 1981; Townroe and Keen 1984); and infrastructure (especially transportation) proliferation coincides with stages of growth, core–periphery, and modernization frameworks (and in a different vein, with political economy analyses that emphasize its control and exploitation aspects).

Donor-nation actions also enter the picture as a corollary of national policies. International loans or technical and financial assistance provide support for government programs; trade and tariff practices of developed nations complement import substitution industrialization or provide favored-nation status for particular economic activities; and basic needs and housing programs of international agencies supplement those of Third World nations. Hence, donor-nation actions also reflect paradigmatic views of development.

To what degree has paradigmatic thinking prompted (or given rise to) policy and donor-nation actions? Alternatively, to what degree has it been a rationalization for actions already taken (or planned)? The issue is interesting, but largely academic. Forces related to national policies and donor-nation actions, also to world economic and political conditions, are a reality of Third World settings – in this book's view, a dominant reality.

Third World landscapes at ground level

Third World landscapes exhibit many elements associated with development constructs. Urban primacy and core–periphery distinctions are the norm; economic activities that many see as modern and traditional stand side by side in dualistic fashion; and dependency relationships are amply evidenced. Lacking, however, is the sense of *process* that comprises the heart of paradigmatic statements – a process driven either by self-sustaining, autonomous growth or by conflictual relations among socioeconomic entities. These forces are present, but not dominant.

Contributing to this realization is the author's earlier work relating migration to development in Costa Rica, summarized in Chapter Three. Initially, development levels were differentiated solely in terms of a rural–urban dichotomy (Brown and Lawson 1985a), a procedure consistent with established views of development. But when focus shifted to specific places that had noticeably high levels of in- or out-migration (Brown and Jones 1985; Brown and Lawson 1985b), it became evident that different factors operated in each locale. Significantly, these factors were artifacts, not of processes depicted by development paradigms, but of forces stressed by the present argument – increased cacao and sugar cane production resulting from price increases in world markets, government-supported expansion of commercial cattle production, land consolidation related to (that and other) government economic initiatives, completion of the Pan American Highway link between Costa Rica and Panama (with donor-nation support), locational shifts in banana production resulting from the introduction of disease resistant varieties, and urbanization itself. Hence, the investigation began with a paradigmatic view of development, but it discovered a different set of forces in searching for *tangible* explanations.

In a similar vein, paradigmatic reasoning provides a rationale for the highly polarized space-economies of Third World settings and related phenomena such as overurbanization and immoderate levels of urban-bound migration. But these outcomes, often summarized by the term urban bias, also have been attributed to exogenous forces highlighted here, particularly to national policies. Recent research finds that exogenous forces provide a more persuasive explanation than does paradigmatic reasoning (e.g., Bradshaw 1985, 1987; Lentnek 1980; Lipton 1976; Taylor 1980, Todaro and Stilkind 1981).

Paradigmatic statements also render a misleading view of development effects on individuals. In populations dominated by the 'just poor' (e.g., Lewis' (1961) *Children of Sanchez*), a television set or ancient automobile represents tremendous improvement over the past. So do more modest gains such as an additional pig or chicken, transportation to the local market, or an outboard motor for fishing; elements of change from 1965 through 1968 in an Ecuadorian coastal village (Thomsen 1969). By contrast, orthodox views suggest transformations of greater magnitude as individuals move from 'traditional' to 'modern', and political economy arguments would denigrate the meagerness of change.[5]

The preceding observations lead to two conclusions concerning change in Third World locales. First, forces stressed by orthodox and political economy approaches are operant, but appropriately seen as components of Third World development, not its totality. Second, exogenous forces highlighted in this book provide a *concrete* explanation of socioeconomic landscape dynamics; an explanation that accounts directly for observed reality and is, therefore, *immediate*.

To further illustrate the perspective advocated here, and provide a means of moving the inquiry forward, consider Heilbroner's (1967: 32) observation that

> despite the flurry of economic planning on top... In the cities,
> a few modern buildings, sometimes brilliantly executed, give
> a deceptive patina of modernity, but once one journeys into the
> immense countryside, the terrible stasis overwhelms all.

The continuing relevance of Heilbroner's observation is illustrated
by Mexico's Yucatan Peninsula, which this author experienced in
1959 and 1985. Change over those twenty-six years is evidenced
by more roads, electrical lines, automobiles, tractors, modern
buildings, government-built worker houses, factories, prosperous-
looking Mexican tourists; also by better integration with other
parts of Mexico, increased life expectancy, decreased infant
mortality, and other artifacts of government and donor-nation
actions – the 'patina of modernity'.

But there is another local landscape wherein a 1950s guide to
shops in Merida (the state capital and major city) is still valid,
rural villages and small towns remain surprisingly familiar, and
little change is evident. That these landscapes exist side by side,
as major ingredients of Yucatan's geography, suggests the *forces*
of change, or development, are *external* and lack the self-sustaining, self-perpetuating energy which is central to paradigmatic
perspectives.

Were there a cessation of development efforts, then, the
Yucatan's landscape might readily revert to an earlier, more
indigenous socioeconomic state, like natural vegetation taking over
a once plowed but now neglected field. In some sense, this has
already occurred. At the century's turn the Yucatan was a major
world producer (often the major producer) of henequen, a plant
fiber used to make rope. But competitive cultivation in other
tropical areas increased; the Mexican Revolution dissolved haciendas, thus eliminating cost savings related to economies of scale;
and later, synthetic fiber substitutes were perfected (James and
Minkel 1986: 41–3; Joseph 1986). As a result of world market
and internal shifts such as these, past prosperity is primarily
evidenced by an avenue in Merida lined by turn-of-the-century,
somewhat tattered mansions, one sees little production of agave
(the henequen plant) in rural areas, and a sense of stasis pervades
the landscape overall.

Relevant to these observations on the Yucatan and accounting
for reality, Wilber and Jameson (1988: 22–5) discuss the importance of *moving beyond paradigmatic thinking*. Of particular
concern is its implicit belief that history is progress, that develop-

ment is an inevitable outcome of time, and that development is an ideal state to be achieved, measured, and studied by social scientists. But history, they note, is convoluted, not linear. For this reason,

> The parable of historical progress common to both the orthodox and political-economy paradigms is a metaphor that may be useful in studying an abstraction...but it is misplaced in studying the **actual** development of Peru or Uganda. (Wilber and Jameson 1988: 22, emphasis added)

In a similar vein, Friedmann and Wulff (1976: 10) note

> [historians] provide a healthy antidote to the facile generalizations of those social scientists who are inclined to think that the start of urbanization in Third World countries coincided more or less with the beginning of their own interest in the study of this process. The historians'...acceptance of the possibility of breakdown, stagnation, and decline of whole civilizations, and his search for the interior meaning of events stands in sharp contrast to the social scientists' preoccupation with universal laws, statistical regularities, and equilibrium.

Yet the question is more than simply history versus social science generalization and a linear concept of development. Observing the United States in 1880, social scientists would not necessarily conclude they were in the midst of a monumental development era; and in 1929, few would conclude they were on the doorstep of major economic depression. Said another way, only with historical perspective do we know with certainty that development has (or has not) occurred, not by being in the middle of events; and only with historical perspective are we able to discern the outlines of operant processes.[6] Accordingly, it may be more relevant that 1880 social scientists focus on the Erie Canal or the burgeoning railroad network as elements of regional or areal change, rather than on development broadly defined.

This is not to say we should eschew the study of development processes, or paradigmatic approaches thereto. Rather, it is time to *augment* that focus by, as Wilber and Jameson (1988: 25, emphasis added) note, "*throwing off the conceptual blinders of the paradigms*", and enlarging thereby the range (and nature) of questions accepted as relevant to understanding Third World development. In this context, the geographer's *sense of place*, whether gained from field experiences or secondary sources, is an exceptionally valuable tool, useful both for evaluating paradigmatic constructs (such as development) and guiding research into new avenues.

Moving beyond paradigmatic thinking

This treatise on Third World development emphasizes world economic conditions, donor-nation actions, and government policies as major agents of change. As exogenous forces that mold the context, or structural conditions, within which local change occurs, they produce the most prominent imprint of development on Third World landscapes.

But there also is an *epistemological* (and methodological) issue to the argument: the need to disencumber our minds by stepping away from both orthodox and political economy views of development. This need is indicated by disparities between paradigmatic portrayals of Third World settings and their actuality. In being there, for example, one does not find individuals and enterprises that are either traditional or modern, as orthodox perspectives suggest; nor does one find capitalist and pre-capitalist endeavors, the political economy equivalent. Distinguishing between indigenous and Western elements is experientially more appropriate, but that also tempts confusion. Consider the dress of Andean Indians which frequently includes an English fedora hat, introduced in the late 1800s, or Otavalan weaving based on designs by Escher, introduced by Peace Corps workers in the mid-1960s. Hence, classifications ought to recognize that Third World landscapes represent an ongoing interpenetration of indigenous and Western elements (Weinstein and McNulty 1980).

This concern with classification is not merely academic. Terms such as modern, traditional, capitalist, and pre-capitalist are value-laden caricatures of economic behavior that misrepresent personal and aggregate qualities. Whether local/individual enterprise leads to development or perpetuates the status quo, its usual intention is to maximize personal or family welfare, and the particular form(s) of this initiative reflects its economic, social, and political context (Peattie 1980).

Further, enterprise that seems traditional (or pre-capitalist) may in fact be a dynamic element of change. In observing agricultural, small-scale, or artisannal economic activities that appear indigenous, one is tempted to apply the word 'traditional', particularly when matched against elements reflecting exogenous forces. But that term implies backwardness, and as Roberts (1977: 183–4) notes,

> The significant aspect about the urban traditional activities... is that hardly any of them are traditional. Indeed, they are as modern in the historical sense as the technologically-based activities which they complement. The danger of describing these activities as traditional, household or marginal lies in giving the impression that they represent outmoded forms of economic activity...[but]...the small enterprise...possesses a flexibility which is a functional component of the current capitalist development process... The small enterprise, not the large, is the active agent of capitalist penetration in Peru... These 'marginal' populations are not passively reacting to economic changes originating from above, but are engaged in their own internal processes of development and in using changes elsewhere in the system to further that development.

Roberts is not alone in seeing 'marginal' undertakings as a major agent of development, although many would dismiss them as traditional. Belsky's (1988) study of the potato marketing system of Ambato, Ecuador provide a highly detailed example of the resourcefulness and vitality of Third World enterprise, as does Peattie's (1980) description of economic activity in Bogota, Colombia, which directly challenges dualistic thinking. Similarly, Yapa (1977: 24) notes

> A common premise is that underdevelopment is related to a lack of enterprise or entrepreneurship... Yet, even the casual tourist has learned to come away with a healthy respect for merchant enterprise in the markets of Third World cities and towns, and the acquisitive skill of the ubiquitous money lender is only too well known.

In light of such evidence, and first-hand experience, fitting reality into categories like modern, traditional, capitalist, or pre-capitalist is inherently suspect. That such terms are expendable is further indicated by their absence in research concerned directly with development alternatives; for example, Freeman and Norcliffe's (1985) study of Kenya's rural-nonfarm sector and the small-scale industry studies of Cortes, Berry, and Ishaq (1987), Little, Mazumdar, and Page (1987), or Pack (1987).

Stepping away from paradigmatic perspectives on development also is warranted by the *dialectic* of academic progress (Arndt 1987). In broad terms, orthodox perspectives have constituted the cornerstone, or 'thesis', of development thinking. Counteracting this through political economy perspectives, the 'antithesis', has required vigorous, often virulent, argument. By now, however, this debate has run its course. Many view orthodox and political economy perspectives as complementary, rather than competing or mutually exclusive explanations, and 'synthesis' is underway.[7]

Armstrong and McGee's (1985) recent study of urbanization in Asia and Latin America illustrates that a shift in orientation is ongoing, even when not identified as such. A political economy framework is set out initially, but subsequently its prominence dims, often to inconsequence, as focus turns to explicating the dynamics of local change. Factors highlighted in this chapter's argument are shown to be important; detailed attention also is given to the social basis of production and consumption as influenced by institutionalized socioeconomic relationships and underlying structural rigidities.

More critically, Armstrong and McGee's highly informative study could have been accomplished *without* reference to political economy constructs. These, as well as orthodox perspectives, have awakened our sensitivity and been important stepping stones to knowledge of Third World development. But the lessons are well known, and embracing or being immersed in either viewpoint is no longer an integral element of inquiry.[8]

That synthesis is underway, and the path it might follow, is more directly shown by the present argument (and the book overall). Its focus on structural aspects of development related to exogenous forces has *roots* in both political economy and orthodox thinking, and *integrating* these approaches is seen as a necessary step towards an appropriate analytical framework. But the argument also observes that intellectual entrenchment of paradigmatic perspectives has obscured their divergence from Third World actualities, in part by blunting the question itself. Required, therefore, is a conscious move away, taking what we have learned from paradigmatic thinking but *disencumbering* research protocols to encourage cross-fertilization from a variety of viewpoints. It is time to broaden our vision. *Generalization* remains the objective; but it needs to *emanate* from, rather than be *imposed* on, the locale being studied.

Development or modernization?

The preceding argument asserts that processes underlying Third World socioeconomic change differ substantially from those posited by conventional or established perspectives. An important element of this contention is the possibility that Third World change does not represent classic development so much as an imposition of contemporary artifacts through the local articulation of world economic and political conditions, donor-nation actions, and policies of Third World governments. This question, Development or Modernization?, is addressed in the present section, leading to A Note on Research Strategy.

Recall that paradigmatic portrayals of development draw heavily (but not exclusively) on North American and Western European experiences (Arndt 1987; Meier and Seers 1984). These nations moved from economies based on primary activity (agriculture and mining), through an industrial revolution, to modernization; a transition broadly characterized as being laissez-faire and driven by internally generated, self-sustaining growth and/or by conflictual relations among socioeconomic entities.[9]

Development involved economic expansion overall, but also massive changes in prevailing economic structures. For example, the United States labor force increased more than six-fold from 1870 to 1970. Accompanying this was a marked decrease in agricultural employment, from approximately 52.5 to 2.9 percent, and an increase in manufacturing, from 19.1 to 33.9 percent. Particular skills also were affected. Flat-boat operators were highly valued by the river transportation industry in the middle 1800s (Mak and Walton 1973), but such employment was rare 100 years later; similarly for blacksmiths. More recent technological change has outmoded machine tool makers and typesetters.

North American and Western European development also involved social change. Economic expansion necessitated rewarding persons on the basis of productivity, ingenuity, and skill, thus lessening the role of social class or the occupation of one's parent. Increasing employment opportunities that spread income throughout society also contributed. Finally, social change was enhanced by growth in educational opportunities. From a societal perspective, these operate to train the workforce and screen out persons less suited for participation in the contemporary economy (Blaug 1973, 1976; Layard and Psacharopoulos 1974; Psacharopoulos and Woodhall 1985), but for individuals, education is a means of self-advancement.

Accordingly, many United States citizens of the early 1900s could look forward to an increasing probability of employment, higher real wages, and upward mobility in occupational and social status, if not in the present then at least in the next generation. But because development is uneven, some prosper while others do not.[10]

That Third World nations will follow a similar scenario is a common assumption in paradigmatic portrayals of development (a *convergence* view). Indeed, a tempered form of the preceding description could be applied to a few Third World settings, and many of its elements are more prevalent.

Nevertheless, as argued in the preceding section, the author believes Third World socioeconomic change has not been driven by internally generated, self-sustaining growth or by conflictual relations among socioeconomic entities, although these elements are present. A more persuasive explanation is that Third World change largely reflects policy and institutional actions oriented towards the introduction (proliferation) of contemporary or 'Western' socioeconomic systems. Said another way, forces prevailing at the present time represent *modernization* rather than paradigmatic-type development and/or the qualitative shifts posited by development paradigms.

One must recognize, however, that the North America–Western Europe experience spanned a century, whereas its Third World equivalent, usually measured from 1950 (Chenery and Syrquin 1975; Morawetz 1977), has been ongoing for less than half that time. Further, broad processes shaping today's Third World may not be discernible to social scientists because, as noted earlier, we see development largely by looking back through history, not while in the midst of events. Finally, to grow economically, Third World nations must find and expand a niche in industrial-technological marketplaces that are already mature, an economic milieu differing considerably from that faced by North America and Western Europe in the nineteenth century. Hence, there may well be an ongoing Third World development process(es), not merely modernization, but one that *deviates* extensively from paradigmatic espousals.

A note on research strategy

Although the perspective outlined above, and this book, is focussed on change in sub-national areal units, rather than broad-
gauged development processes, these are interrelated; and expanding our knowledge in either domain would be facilitated by a *shift*
in research strategy. This involves moving away from inquiry
organized around established development frameworks; for example, demonstrating a fit between paradigmatic expectations and
selected Third World elements.

Better suited to the task ahead is a *less nomothetic, more
idiographic* approach that takes account of places as entities in
their own right and their experience, or perception, of change;
what might be seen as a behavioral geography of places. Third
World settings are the fount from which research frameworks
should emanate. Selected elements of established (orthodox and
political economy) perspectives might be included, but only if
significant to and warranted by the situation being considered.
Essential, however, is an ongoing dialogue with place *reality*, and
continual revision to align one's framework accordingly. The
ultimate goal is a better understanding of development. But while
we await that achievement, the proposed research tack insures a
dramatic increase in knowledge of the Third World at *ground
level*, a major gain in itself.

Discontent with customary research strategies stems from the
author's frustration in meshing an experiential sense of the Third
World with its social science portrayal. Interestingly, his disquiet
has been echoed by others with dramatically different research
concerns; and they also urge a shift away from strongly nomothetic objectives. Examples include Anselin's (1988: ch. 9) focus on
'spatial heterogeneity', Golledge's (1988) on 'multiple realities',
Sayer's (1984, 1985) on 'realism', and more close to the concern
here, 'locality' studies in Britain (Cooke 1987; Jonas 1988; Massey
1983, 1984: ch. 5).

This issue will be returned to in the concluding chapter, Eight.
Leading up to that, however, are studies which *incrementally*
provide evidence in support of the development perspective
delineated above. These studies also portray an evolution in
understanding, or comprehending, development. Hence, the
account in Chapter Three of aggregate migration flows in Costa
Rica represents a research strategy which is routine in the sense
that it follows closely on approaches taken by many others. By

Chapters Six and Seven, however, discourse progresses to regional change in Ecuador under a research scheme better aligned with the author's current view of Third World development.

3 Aggregate migration flows and development, with a Costa Rican example

The preceding chapter first presented a synopsis of established development frameworks; then elaborated the book's framework centered on world economic and political conditions, donor-nation actions, and policies of Third World governments. These positions represent a conceptual transition that is mirrored in studies of aggregate migration flows within Third World nations.

To elaborate, Third World migration studies have traditionally employed research strategies that embody neoclassical precepts, align with dual economy and human resource views of development, and are nomothetic in the sense that they promote a universal model. With the introduction of political economy perspectives to migration studies, focus shifted towards the role of socioeconomic structure; and more or less simultaneously, others (e.g., the author) argued that development should be an explicit, rather than background, variable. Subsequent empirical analyses to delineate socioeconomic structure and development influences on migration highlighted the importance of localized, place-specific, relatively unique occurrences, i.e., idiographic elements. An amalgamation of these findings spawned the author's perspective on development, rather than corroborating established conceptualizations, and indicated a need for research strategies with an idiographic dimension, focussed on place contexts giving rise to migration.

Chapter Three reflects the initial phases of the author's involvement in this transition. Research recounted here was initiated in the early 1980s when debates between neoclassical and political economy approaches to development were rampant, and the dialectic of academic progress, referred to earlier, was in midstream. Then, the author's primary concern was migration. His plan was to articulate its conceptual link with development and implement statistical analyses wherein contrasting views of development would be operationalized and their relative importance evaluated. But operationalization was diverted because variables uniquely representing different development frameworks could not be identified, and the frameworks themselves did not coincide with the author's sense of Third World reality. As a result, development was treated largely as a 'black box' phenomenon.

As an entry point to migration studies, this chapter first considers statistical formulations rooted in economics which exemplify the nomothetic search for laws that apply across geographic settings and points in time. Referred to here as the *conventional model*, it was initially articulated for Developed World settings such as the United States, but applied with minimal modification to Third World nations which are highly diverse in development characteristics. Further, just as cross-national variation is disregarded in formulating those models, so are development differences among sub-national political units.

After briefly reviewing the conventional model and its applications, attention turns to elaborating the theme that Third World migration processes are affected by spatial variation in place characteristics related to development. This is initially demonstrated by reference to existing literature; then by a conceptual framework termed a development paradigm of migration. Next, the conventional model is applied to aggregate migration flows in Costa Rica in a manner demonstrating that its coefficients vary locationally, and that these variations relate to events identified with development. Finally, attention is given to the issue of variables other than those in the conventional model and specificity of explanation or understanding. A summary and concluding observations completes the chapter.

Conventional modeling of migration in Third World settings

Migration has been most commonly studied under economic conceptualizations that make no explicit reference to development. One variant of this is the *labor force adjustment model*, which sees migration as an individual response to place differences in wage rates and job opportunities at a single point in time, and as an equilibrating mechanism for eliminating such differentials by matching labor supply and demand. More specifically, aggregate migration from i to j would be enhanced if the latter had a higher wage rate, more employment opportunities, and these differentials were communicated to place i (Lowry 1966). Applications of this model in Third World settings include Greenwood's studies on

Egypt (1969), India (1971a, 1971b), and Mexico (Greenwood, Ladman, and Siegel 1981); Beals, Levy, and Moses (1967) on Ghana; Carvajal and Geithman (1974) on Costa Rica; Falaris (1979) on Peru; Levy and Wadycki (1972, 1974a, 1974b) on Venezuela; and just recently, Barber and Milne (1988) on Kenya and Gottschang (1987) on Manchuria.

The labor force adjustment model takes an aggregate, systems view of migration, but embodies the behavioral assumption that individuals respond to wage rate or employment opportunity differentials. Because this model also assumes free mobility of labor, individual responses are seen to be immediate, that is, without regard to inertia effects at the origin or future conditions at the origin and destination.

The *human capital*, *cost benefit*, or *expected income* approach also assumes individuals respond to wage rate or employment opportunity differentials, but sees this in terms of likely future conditions as well as present ones. Although Sjaastad (1962) was an early proponent, Todaro (1969, 1976) pioneered this framework in Third World settings. Todaro's model states migration will occur if the expected present value of future earnings in the potential destination exceeds that of the origin, which is dependent on the costs of migration, place-to-place differences in income levels over the migrant's planning horizon (years for which future earnings are considered), and the probability of securing employment in each place. Thus, immediate income at a potential destination may be less than at the origin, but if the stream of expected income over time is greater, migration should occur.[1]

Applications of the human capital model in Third World settings include Godfrey (1973) on Ghana, House and Rempel (1980) and Rempel (1980, 1981) on Kenya, and Speare (1971) on Taiwan. More common are studies that conceptualize migration in human capital terms, but employ empirical analyses similar to those emanating from the labor force adjustment framework. Examples include King (1978) on Mexico, Fields (1979, 1982) and Schultz (1971) on Colombia, and Schultz (1982) on Venezuela. One reason for this discrepancy may be that available, usually aggregate, data generally are not sufficiently rich to estimate future earnings, much less the migrant's perception of future earnings. Without that, human capital and labor force adjustment models are not distinguishable.

Although conventional models emphasize spatial variation in wage levels and job opportunities, an equally important component

is information flows between places, which communicate economic differentials to prospective migrants. Gravity model variables
or distance between origin and destination are commonly used as
surrogates (see migration references above). Given concern for
development influences, however, particular note should be made
of the *migration chain* effect, i.e., that current migrants tend to
follow the paths of relatives, friends, and acquaintances who have
moved earlier. The strong role of migration chains in Third
World settings reflects a paucity of formal communication mechanisms; significance of one's community, which enhances the
perceived reliability of informal communications sent through
migration chains; and the role of friends, relatives, and acquaintances in a destination who ameliorate adjustment should migration
occur.[2] In terms of relative importance, this variable usually is
statistically stronger than wages, job opportunities, and other
economic factors, although the disparity would be a development-
dependent balance.[3]

Development–migration relationships
in previous research

Development is not explicit in conventional models of migration,
but it is present nevertheless. First, market conditions, which
occupy a central role, vary from place to place in reflection of
economic growth and decline experiences at national, regional,
and local levels. Second, the way individuals respond to market
conditions is itself an artifact of development. Finally, wage rate
and job opportunity differentials are communicated to prospective
migrants through information flows wherein informal communications via migration chains outweigh media influences (Levy and
Wadycki 1973; Rempel 1980, 1981), but as noted above, this
balance is development-dependent.

For an explicit consideration of the development–migration
interface, one may turn to dual economy conceptualizations,
reviewed in Chapter Two. In these, movements from traditional
to modern locales are motivated by wage and job opportunity
differentials between them; migration persists until the modern
sector develops sufficiently to transform the traditional and/or
absorb its excess labor; and meanwhile, economic growth proceeds
by drawing on extensive human resources provided by traditional

areas. This dynamic is further elaborated in spatial forms of the dual economy model, notably core–periphery and growth center constructs. These distinguish between polarization or backwash effects, whereby economic growth impulses move toward the core or growth center(s), and spread or trickle-down effects, whereby growth impulses are directed to places in the periphery. The balance between these effects determines whether economic conditions, represented in conventional migration models by wage and job opportunity variables, are more attractive in core or periphery locales, and migration flows would shift accordingly.

Complementing dual economy, core–periphery, and growth center constructs is the human resource perspective, one segment of which addresses migration effects on development (Brown and Kodras 1987; Brown and Lawson 1989; Gober-Meyers 1978a; Myrdal 1957). Because migration is selective, origin places are drained of quality human capital to the benefit of destinations, thus altering development prospects of each locale. To put this in a broader framework, migration occurs in response to place-specific economic differentials between core and periphery (growth center and hinterland, modern and traditional sectors), which then are exacerbated by human resource effects on development trajectories, leading to further migration, and so on, until polarization reversal (i.e., the switch away from core dominance).[4]

Hence, research in the tradition of neoclassical economics suggests a highly interrelated system wherein migration and the development context within which it occurs have significant effects on one another. Nevertheless, studies of aggregate migrations give minimal attention to place characteristics associated with development. Prevailing instead are countrywide analyses employing a variant of either labor force adjustment or human capital conceptualizations.

By contrast, development is a primary concern of historical-structural perspectives, a variant of political economy approaches reviewed in Chapter Two. These seek generalizations concerning forces dictating the political, economic, social, and geographic organization of society; and migration is viewed as resulting from those forces interacting with local conditions (Bach and Schraml 1982; Balan 1983; Gregory and Piche 1981; Riddell 1981; Roberts 1978: esp. ch. 4; Swindell 1979; Wood 1982). Said another way,

generalizations do not take the form of statements about...empirical patterns but rather...[concern] abstracted tendencies and mechanisms which produce specific...patterns under conditions and circumstances which are contingent and historically specific. (Holmes 1983: 255)

Accordingly, conclusions concerning migration pattern and process tend to be stated in terms of the locale being studied, not as universals, and migration may even be regarded as a purely derived phenomenon. For example, Friedmann and Wulff (1976: 26–7) observe

migration to cities reflects merely a demographic adjustment to changes in the spatial structure of economic and social opportunities... [It] is a derived phenomenon, a symptom of urbanization and not the thing itself... Demographers and others insisted on treating migration as a major policy variable when it was, in fact, dependent on the major structural features of the economy.

One manifestation of development to which historical-structural research has drawn attention is the role of origin factors in fomenting migration. Land availability and income-producing opportunities in rural areas, for example, are affected by development-related phenomena such as high rates of population increase, technology diffusion, urban bias in the location of manufacturing, and prevailing social-political structures. Illustrative in this regard is Costa Rica's promotion of commercial cattle production (Taylor 1980). Because such production is markedly less labor-intensive than alternative agricultural enterprises and in Costa Rica involved land reform measures that disrupted traditional subsistence agriculture, income-producing opportunities for the peasant were reduced and out-migration increased. This exemplifies, more generally, Brown and Lawson's (1985b: 417) assertion that

structural change or disequilibrium of any sort, including cataclysmic events...may be the single most important progenitor of population movements in Third World settings.

Further, Taylor (1980: 86) notes that the importance of origin factors

contradict(s) the traditional view that urban attraction is at the root of the migration process... [Data] show that most peasants leaving Guanacaste have favored destinations within the periphery over the urban center. At the same time...the restriction of employment and land owning opportunities within the periphery make it increasingly difficult for the peasant to avoid a final move into the city. It seems that urban pull factors are playing a relatively minor role in the Costa Rican rural–urban migration process.

Also relevant to Taylor's point is evidence of sizeable migrations to rural destinations in a number of Third World countries (Brown and Lawson 1985b), which is counterintuitive to neoclassical, pull-oriented models of development.

Although research has tended to take either a neoclassical or historical-structural stance, the *complementarity* of these views suggests a need for integrating within one framework structural-institutional forces and individual or household responses to such factors. The issue is addressed in this and the following chapters, but particularly in the next section where a 'development paradigm of migration' is set out. This melds macroeconomic structure and individual decision-making by emphasizing the importance of regional character or place characteristics related to development.

Structural forces and individual decision-making also have been integrated by using simultaneous equations or two-stage least squares formulations that explicitly treat the migration–development link (Gober-Meyers 1978a, 1978b; Greenwood 1975a, 1975b; Okun 1968; Salvatore 1981). However, this approach has rarely been applied to Third World settings, Greenwood's (1978) study of Mexico being one exception. More important, by estimating parameters on a countrywide basis simultaneous equations models embodying migration–development links have assumed that similar relationships govern all locales, whereas Brown and Sanders (1981) argue that the role of migration process components may differ considerably across development milieus. Also, simultaneous equations models generally measure development in terms of a single variable whereas a preference for multidimensional scales is indicated in Chapter Four of this book.

A development paradigm of migration

This section outlines a conceptual framework based on the

observation that economic growth involves structural changes in society which, in turn, alter the role of factors influencing migration. Extending earlier work by Mabogunje (1970), Zelinsky (1971), and Brown and Sanders (1981), it argues that conditions pertinent to human movements are affected by the development-dependent mix of social and economic conditions, government policies, infrastructure, technological achievement, and other aspects of regional systems. Accordingly, modern sector wage rates and job opportunities play a dominant role in advanced settings, whereas migration chain or rural push effects dominate under less advanced conditions. Different development milieus give rise, therefore, to different 'processes' of migration. Further, since development varies both temporally and from place to place, these differences are found for a nation at different times in its evolution or, alternatively, in comparing nations or sub-national regions at a given point in time. Given this, ambiguities in the findings of previous migration research may be attributed to differences in the level or nature of development among the locales studied (Jones and Brown 1985) or to differences in the (historical) developmental processes characterizing those locales.

To elaborate this perspective, attention turns to migration directionality, the relative role of factors motivating or guiding migration, and changes in both over the course of development. These are examined in terms of (1) the locus, range, and mix of job opportunities; (2) the degree of origin push versus modern sector pull; (3) social system characteristics; and (4) the proliferation of transportation and communication infrastructures. Development is depicted through a combination of dual society and stage precepts, which are employed solely because they provide an uncomplicated medium for illustrating the *link* between development and migration, the *primary concern* of this exercise. However, because societal structure and its change is a shared focus among development constructs, using a different construct should lead to a similar conclusion – that the socioeconomic milieu in which migration occurs is a critical element and affects the role of migration factors accordingly.

In traditional society, the locus, range, and mix of job opportunities are relatively undifferentiated across the landscape and would not, therefore, induce a significant amount of permanent migration. Over time, however, labor market artifacts of a more contemporary society are established, first in larger cities and later in intermediate and smaller ones. Prototypically, this begins by interjecting modern sector activity into the more traditional small-scale enterprise economy, leading to gradual transformation of the latter and overall growth. The aggregate number and range of job opportunities thus increase as development progresses. There also is a change in the ratio of formal to informal/small-scale enterprise employment. Specifically, the relevance of modern sector job opportunities and wage differentials increases over time and spreads spatially. Concomitantly, the role of informal and small-scale enterprise sectors decreases as they mesh with and are absorbed by the modern and formal. In terms of present actualities, urban systems of the Third World remain primate or highly focal in nature, informal and small-scale enterprise sectors dominate, integration of urban and rural/small-town economies is fragmentary, and migration patterns are oriented towards the largest city(ies). However, where modern sector activities have noticeably diffused to intermediate-size cities, as in Mexico or Venezuela, contemporary migration patterns exhibit a parallel decentralization (Betancourt 1978; Chaves 1973; Chen 1978; Chen and Picouet 1979; Greenwood, Ladman, and Siegel 1981; Scott 1982).

With regard to origin push, the initial rural situation is subsistence agriculture and a balance between population and available resources. An early artifact of modernization has been the diffusion of health-related innovations, leading to a fall in death rates. As a result, more children survive to adulthood, thus increasing (considerably) population pressure on the land and on existing agricultural systems.[5] One response has been migration or circulation. Another has been to alter agricultural production systems through innovation, a response heavily promoted by domestic and foreign forces external to rural communities. But the end result is similar because innovation diffusion generally leads to exacerbating social and economic disparities within the community and/or to increasing the superfluity of labor, both root causes of migration (Brown 1981: ch. 8; Connell, Dasgupta, Laishley, and Lipton 1976: 200; Gotsch 1972; Lipton 1980; Saint and Goldsmith 1980). Partly in response to this situation, there has been exhortation towards diffusing appropriate, labor-biased technology to rural areas, redistributing land, and other social structure directed measures. Presently however, rural population pressure remains high, and until development brings a better balance between core and periphery, origin push should be a significant migration factor (Connell, Dasgupta, Laishley, and

Lipton 1976; Rhoda 1979, 1983; Silvers and Crosson 1983).

Social system characteristics, as they affect migration, and the proliferation of transportation and communication infrastructures are interrelated. Initially, when infrastructures are sparse, inter-personal contacts among family and acquaintances are the primary source of information, assimilation at destinations is highly dependent on earlier migrants, and accordingly, migration chains are well defined. As infrastructures proliferate, a number of changes occur. First, information from other than interpersonal sources is more readily available. Second, because movement costs are lower, circulation strategies are more common; this enables potential migrants to gradually acclimate themselves to destinations, thus reducing reliance on previous migrants. Third, increased contact between modern and traditional segments of society alters value systems and, particularly, social norms pertinent to migration. Finally, proliferation of transportation and communication infrastructures makes more feasible (both economi-cally and socially) the spread of economic activity from larger cities to those of secondary importance, an occurrence which alters the locus, range, and mix of job opportunities and related migration patterns.

Earlier attempts to link development and migration primarily focussed on movement patterns, not processes.[6] Connell, Dasgup-ta, Laishley, and Lipton (1976: 201), for example, conclude that

> Patterns of migration from a rural community may well change in stages, following integration [into the national urban system] and development of that community. Circular migration usually comes early...succeeded by directed migration, but still relatively little differentiated by socioeconomic group... Subsequent integration often differentiates migrant streams... both by status and by age, sex, and destination... The process also often involves a shift from personal to household migra-tion.

Similarly, Zelinsky's (1971) *hypothesis of the mobility transition* posits rate and pattern changes in terms of five development phases: A Premodern Traditional Society, An Early Transitional Society, A Late Transitional Society, An Advanced Society, and A Future Superadvanced Society (Figure 3.1). Skeldon (1977) applies Zelinsky's framework to an area of highland Peru (Cuzco) for its period of change from a premodern traditional society to

Phases of the Mobility Transition

PHASE I—*The Premodern Traditional Society*
(1) Little genuine residential migration and only such limited circulation as is sanctioned by customary practice in land utilization, social visits, commerce, warfare, or religious observances

PHASE II—*The Early Transitional Society*
(1) Massive movement from countryside to cities, old and new
(2) Significant movement of rural folk to colonization frontiers, if land suitable for pioneering is available within country
(3) Major outflows of emigrants to available and attractive foreign destinations
(4) Under certain circumstances, a small, but significant, immigration of skilled workers, technicians, and professionals from more advanced parts of the world
(5) Significant growth in various kinds of circulation

PHASE III—*The Late Transitional Society*
(1) Slackening, but still major, movement from countryside to city
(2) Lessening flow of migrants to colonization frontiers
(3) Emigration on the decline or may have ceased altogether
(4) Further increases in circulation, with growth in structural complexity

PHASE IV—*The Advanced Society*
(1) Residential mobility has leveled off and oscillates at a high level
(2) Movement from countryside to city continues but is further reduced in absolute and relative terms
(3) Vigorous movement of migrants from city to city and within individual urban agglomerations
(4) If a settlement frontier has persisted, it is now stagnant or actually retreating
(5) Significant net immigration of unskilled and semiskilled workers from relatively underdeveloped lands
(6) There may be a significant international migration or circulation of skilled and professional persons, but direction and volume of flow depend on specific conditions
(7) Vigorous accelerating circulation, particularly the economic and pleasure-oriented, but other varieties as well

PHASE V—*A Future Superadvanced Society*
(1) There may be a decline in level of residential migration and a deceleration in some forms of circulation as better communication and delivery systems are instituted
(2) Nearly all residential migration may be of the interurban and intraurban variety
(3) Some further immigration of relatively unskilled labor from less developed areas is possible
(4) Further acceleration in some current forms of circulation and perhaps the inception of new forms
(5) Strict political control of internal as well as international movements may be imposed

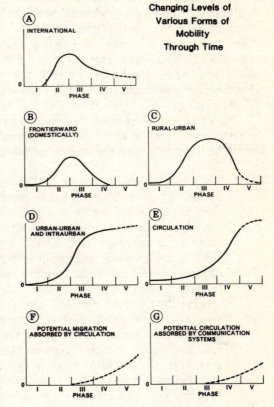

Changing Levels of
Various Forms of
Mobility
Through Time

Figure 3.1 Zelinsky's hypothesis of the mobility transition

an early transitional society, finding that pattern and rate shifts diffuse down the settlement hierarchy and from richer to poorer social groups within each settlement. Underlying this pattern is the accessibility (or distance) of each settlement to major urban centers and its resource endowment; similarly for social groups.[7] Forbes (1981) notes the parallel between his findings for Indonesia and Zelinsky's hypotheses, but argues that socioeconomic structures and their articulation with local conditions differ from place to place in a manner that defies generalization, essentially taking a historical-structural position. Finally, Goldstein (1981) and Urzua (1981) note the importance of incorporating the development dimension if we are to understand migration in Third World settings.

The preceding discussion may be summarized by a three-stage schema, explicated in terms of conventional model variables and shifts in their role over the course of development (Figure 3.2).

- Early migrations, occurring in Zelinsky's early transitional society or, as in Figure 3.2, the move towards modernization, are hypothesized to be highly chain in nature, origin-pushed, and oriented towards opportunities of the informal, small-scale enterprise labor market. Rural-to-rural migration streams are as likely as rural-to-urban ones.

- As development proceeds, entering later phases of the move towards modernization or Zelinsky's late transitional society, migration by more well-off social classes is pulled by educational and modern sector employment opportunities, but retain a significant chain dimension owing to transportation and communication systems that are somewhat rudimentary. At the same time, migration by less well-off social classes maintains its origin-push motivation, orientation towards the informal, small-scale enterprise labor market, and chain characteristics. Rural-to-urban flows increase.

- Finally, as development reaches a relatively advanced level, entering modernization or Zelinsky's advanced society, migration of all social classes is oriented towards formal, modern sector employment and formal communication channels are the primary sources of information, thus reducing and in many instances eliminating the chain dimension. The dominant pattern of migration is urban-to-urban, rather than rural-to-urban.

Development Phases		Factors of Migration						
Brown and Sanders (1981)	Zelinsky (1971)	Wages and/or Job Opportunity Differentials		Education and other Amenity Pulls	Origin Push	Migration Chain Effects	Formal Communication Channel Effects	Pattern of Migration
		Modern Sector	Informal and/or Small Scale Enterprise Sector					
Early Move Towards Modernization	Early Transitional Society	Affects only a Small Segment of the Population	Significant for all Social Classes	Affects only a Small Segment of the Population	Significant for all Social Classes	Significant for all Social Classes	Affects only a Small Segment of the Population	Rural to Rural and Rural to Urban
Later Move Towards Modernization	Late Transitional Society	Significant for the more well off Social Classes	Significant for the less well off Social Classes	Significant for the more well off Social Classes	Significant for the less well off Social Classes	Significant for all Social Classes but moreso for the less well off	Somewhat Significant for the more well off Social Classes	Increase in Rural to Urban
Modernization	Advanced Society	Significant for all Social Classes	Affects only a Small Segment of the Population	Significant for all Social Classes	Significant for the less well off Social Classes but Trend towards affecting only a Small Segment of the Population	Significant for the less well off Social Classes but Trend towards affecting only a Small Segment of the Population	Significant for all Social Classes	Urban to Urban
Trend of Each Factor's Role in Migration Over the Development Process		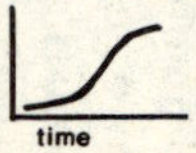 time	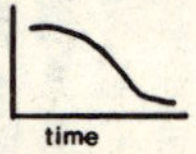 time	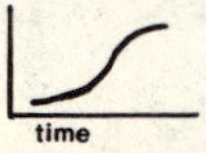 time	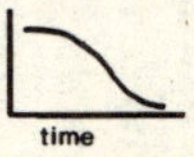 time	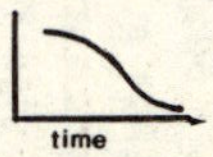 time	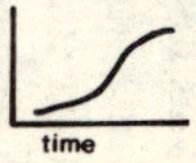 time	
Preponderance of Each Factor in Society at Large		These graphs would parallel the above						

Figure 3.2 Shifts in the role of migration factors over the course of development

Two further observations are appropriate.

First, the above schema is directed towards aggregate migration flows (paralleling conventional models) and the effect on those of place characteristics related to development. Also relevant is migration at the individual level, the operation of personal attributes therein, and how these are affected by place characteristics. These topics are addressed in Chapter Four.

Second, in examining aggregate migration flows, places constituting observations are likely to represent different development milieus. Locales in the periphery, for example, might be 'moving towards modernization', while ones in the core are 'entering modernization'; and as indicated in Figure 2.2, further differentiation also is common. Hence, each set may be characterized by different migration processes. This is illustrated in the following section.

Intercantonal migration in Costa Rica and development milieu effects[8]

One means of addressing place-to-place variation in the importance of factors influencing migration is to estimate separate migration equations for sets of places, where each set represents a different development milieu. An example is Brown and Lawson (1985a), which considers urban- and rural-directed/originated migration streams of Costa Rica for the period 1968–73.[9] This verified a number of propositions derived from the development paradigm of migration framework, elaborated above. In particular, given that urban settings represent more completely articulated market economies, or core regions of development, conventional modeling describes urban-based migrations better than rural ones, as indicated by overall levels of explained variance. Second, variables closely associated with the market economy had a lesser role in rural settings (as indicated by zero-order correlations, betas, and the like), but the friction of distance was greater because of deficiencies in infrastructure articulation. Third, push factors of migration compared to pulls in importance, especially in rural locales. Finally, variables other than those linked with the market economy had greater effect in less advanced places, thus providing an element differentiating models of urban- and rural-based migration.

An alternative analytical approach begins by estimating a single model with *spatially varying parameters (SVPs)* to indicate locational variation in the role of each independent variable and provide an integrated spatial portrait of the development–migration link. This could be accomplished by adding X–Y coordinate variables to a conventional model, and combining them with other variables to form interaction terms.[10] Here, however, two SVP models are estimated: one on the probability of out-migration from i; a second on the probability of choosing j, given that leaving i has been decided. Thus, for reasons elaborated in the following sub-section, separate treatment is given to components of the migration rate, the usual dependent variable in conventional modeling.

Having estimated SVP models, attention turns to identifying place-specific (geographic) occurrences that account for shifts in the role of conventional variables. Identifying migration–development links in this manner may be seen as an application of Casetti's (1972, 1982, 1986) expansion method paradigm.

The spatially varying parameter approach is applied to inter-cantonal migration in Costa Rica for the period 1968–73. Modeling procedures are first presented, followed by comments on interpretation of variables and expected relationships. Attention then turns to empirical findings, both without and with spatially varying parameters. A discussion concludes the section.

Modeling procedures

Spatial variation in conventional model parameters is identified through statistical procedures outlined by Jones (1983, 1984a, 1984b), which are an adaptation of Casetti's (1972, 1982) method for systematically expanding initial regression models. Specifically, the b-coefficients of regression are redefined as functions of X–Y coordinates indicating the geographic location of each observation. Thus, to obtain a second order polynomial surface reflecting spatial variation in the importance of an independent variable W, one estimates

$$Z = a + b'W \qquad (3.1)$$

where Z is the dependent variable and b' the spatially varying parameter specified as

$$b' = b_0 + b_1X + b_2Y + b_3XY + b_4X^2 + b_5Y^2 \qquad (3.2)$$

The full model, then, is

$$Z = a + (b_0 + b_1X + b_2Y + b_3XY + b_4X^2 + b_5Y^2) \, W$$
$$= a + b_0W + b_1XW + b_2YW + b_3XYW + b_4X^2W + b_5Y^2W \quad (3.3)$$

where Z, W, X, and Y are place-specific attributes. After estimating (3.3), W's parameter for each location is computed by placing significant values of b_0 through b_5 in equation (3.2) with appropriate values of X and Y; the resulting set of spatially varying parameters is then mapped. Higher-order surfaces also can be generated.[11]

In applying this approach to human movements, the most obvious dependent variable is the migration rate (Shryock and Siegel 1976: 387–90), defined as

$$m_{ij} = \frac{M_{ij}}{P_i} \qquad (3.4)$$

where M_{ij} is the number of persons migrating from place i to place j in a given time interval t to t+1, and P_i is the population of origin i at time t. Thus, m_{ij} is the percentage of population at risk in place i at time t who move to place j by time t+1 or, in a stochastic framework, the probability that a person in i moves to j in the interval t to t+1.

But the above rate is a composite of two others (Brown and Lawson 1985a), which are more useful for issues addressed in this chapter. One component is the probability of out-migrating or finding a new residence (anywhere),

$$1 - \frac{M_{ii}}{P_i} \qquad (3.5)$$

where M_{ii} is the number of persons remaining in i during the interval t to t+1 and P_i is as above. The second migration rate component,

$$\frac{M_{ij}}{\sum_{k=1}^{n} M_{ik}}$$

(3.6)

for all j and k not equal to i,

is the probability of relocation to j, given that migration has been decided upon. Hence, the migration rate between i and j (term (3.4)) is actually the joint probability that a potential migrant in i decides to seek a new residence (term (3.5)) and, having done so, chooses to relocate in j (term (3.6)).

Each dependent variable conforms with a phase of Brown and Moore's (1970) migration conceptualization, which has been extended to Third World settings by Brown and Sanders (1981: 150–3).[12] This approach has the further advantage of highlighting the role of origin push and related factors, which are especially significant for migration in Third World settings, but are masked by migration rate models.[13]

To operationalize the dependent variables, M_{ij} is the number of persons in canton j in 1973 who are five years of age or older and resided in canton i in 1968; P_i is the population of canton i in 1968 who survive to 1973.[14] Using Z as a generalized dependent variable to indicate either the probability of out-migration from canton i, the probability of relocating to canton j, or the migration rate, the model to be estimated is

$$Z = a + b'(DIST_{ij}, POP_j, WAG_i, WAG_j, PCUJOB_i,$$
$$PCUJOB_j, PRES_i, PRES_j) \quad (3.7)$$

where $DIST_{ij}$ is the distance in kilometers between the population centroids of cantones i and j; POP_j is the population of canton j in 1973; $WAG_{i(j)}$ is the average monthly wage per capita (in Colones) for canton i(j) in 1973; $PCUJOB_{i(j)}$ is the percent of urban-based employment for canton i(j) in 1973, that is, secondary and tertiary sector employment as a percent of total employment; $PRES_{i(j)}$ is the population pressure for canton i(j) in 1973, measured as its total population divided by the number of persons employed in the primary, secondary, and tertiary (all) sectors taken together; and b' indicates that the function is calibrated in terms of spatially varying b-coefficients specific to each independent variable.

If migration rate (term (3.4)) is the dependent variable, all independent variables of (3.7) are used. Out-migration (term (3.5)) is estimated only on origin (i) variables; relocation (term (3.6)) only on independent variables pertaining to destinations (j) and $DIST_{ij}$. In statistically estimating these various models all variables are transformed to logarithms, following routine practices.[15]

Substantive interpretation and expected relationships

As is typical of conventional modeling, the above sees migration as a response to place-to-place differentials in wage rates and employment opportunities, and the degree those are communicated between i and j. More specifically, the population size (POP_j) and distance ($DIST_{ij}$) variables of (3.7) comprise a gravity model component wherein migration between two places varies directly with their size and inversely with the distance separating them. These relationships often are seen as representing information flows between locations, following Isard (1960: 67–79, 493–568), but other interpretations are possible if variables are considered individually. Destination population, for example, can indicate employment opportunities, particularly those of the informal sector, or other amenities. Under this interpretation, POP_j relates directly to in-migration, but an inverse relationship would be anticipated under Alonso's (1977) argument that congestion in populous destinations deters such flows. Likewise, distance may be a measure of financial and psychic costs related to moving, or of transportation system efficiency (modernity), and in both interpretations, $DIST_{ij}$ and the i-to-j migration flow should relate inversely.

Expectations related to the wage, urban job, and population pressure variables assume that people move from low- to high-wage areas and from fewer to more numerous employment opportunities. Hence, WAG_i and $PCUJOB_i$ should vary inversely with out-migration, while in-migration should vary directly with WAG_j and $PCUJOB_j$. Parallel but opposite relationships are expected for the ratio of population to total employment ($PRES_{i(j)}$), an indicator of pressure on an area's economic base and its likelihood of providing sufficient support for local residents. Thus, $PRES_i$ should vary directly with out-migration, and $PRES_j$ inversely with in-migration.

More central to this chapter are spatial differentials in the above relationships, i.e., that migration factors should operate differently as a consequence of local conditions. Three considerations enter into expectations. First, core–periphery or two-sector growth formulations suggest a spatial order to development; in the present instance, centering on the San Jose/Meseta Central core and declining with distance from those locales. Second, while this picture reasonably corresponds to the character of economic activity in Costa Rica, there also are pockets of modernity in the periphery, primarily as commercial agriculture. Finally, standard expectations, those outlined in earlier paragraphs of this sub-section, derive from research with an urban or modernity bias and should, therefore, prevail in core areas.[16] For places where traditional or informal economic systems are more prevalent, however, previous research provides little insight; hence relationships could be stronger, weaker, similar, or even opposite.

To summarize implications of these considerations, the role of independent variables should vary spatially in accordance with core–periphery patterns, but with deviations reflecting the location of commercial agriculture, development projects, and the like. Further, standard relationships should typify core areas, whereas other, and possibly opposite ones should hold in peripheral locales or development nodes. To further explore these contentions, attention turns to migration in Costa Rica for the period 1968–73. Focus is first on the conventional model; then on its implementation with spatially varying parameters.

Empirical findings: application of the conventional model

Table 3.1 presents statistical estimates of the out-migration, relocation, and migration rate models, the latter to enable comparison with earlier studies. The out-migration analysis treats each canton as an observation, yielding an n of 59; the other two analyses employ as observations all possible dyads among the cantones, less the 59 i-i dyads, yielding an n of 3422.

Turning first to out-migration, this is inversely related to the proportion of urban employment in a canton ($PCUJOB_i$) and directly related to its level of population pressure ($PRES_i$), but elasticity estimates indicate out-migration is more responsive to a percentage change in $PRES_i$ than in $PCUJOB_i$ (b=+1.62 versus -0.43, respectively). That WAG_i is not significant suggests out-

Table 3.1 Conventional models of migration: statistical estimates for Costa Rica, 1968–73

Independent variables[a]	Out-migration analysis				Relocation analysis				Migration rate analysis			
	r	b	Beta	t	r	b	Beta	t	r	b	Beta	t
DIST-ij					-0.26	-0.65	-0.35	-26.48*	-0.20	-0.59	-0.35	-21.40*
POP-j					0.56	0.79	0.44	28.55*	0.43	0.60	0.37	20.89*
WAG-i	-0.36	0.30	0.15	0.98					-0.13	0.61	0.08	3.89*
WAG-j					0.48	2.38	0.30	13.85*	0.35	1.57	0.21	8.79*
PCUJOB-i	-0.55	-0.43	-0.53	-3.68*					-0.11	-0.62	-0.20	-10.11*
PCUJOB-j					0.24	-0.67	-0.20	-11.33*	0.17	-0.60	-0.20	-9.74*
PRES-i	0.43	1.62	0.29	2.16*					0.12	2.33	0.11	6.21*
PRES-j					-0.38	-2.58	-0.11	-7.11*	-0.30	-2.09	-0.10	-5.58*
	n = 59				n = 3422				n = 3422			
	r² = 0.36				r² = 0.47				r² = 0.33			

Note

(a) See text for definitions of dependent and independent variables.

* indicates that variable is significant at the 0.05 level or better.

migration is motivated by a deficit in employment opportunities rather than low wages. This is consistent with the widespread reliance (in lesser developed economies) on non-monetary sources of income such as payments-in-kind, supplementary subsistence agriculture, or apprenticeship systems related to rural-nonfarm activity (Lipton 1976; Schneider-Sliwa and Brown 1986).

In relocation, on the other hand, place-specific wage rates play a significant role; other relationships indicate that destination cantones are more populous (POP_j), closer to the origin canton ($DIST_{ij}$), and better endowed with employment opportunities per capita ($PRES_j$). Only the proportion of urban-based employment ($PCUJOB_j$) does not behave as expected, with an inverse, rather than direct, relationship to in-migration when evaluated in concert with other independent variables. Finally, considering relocation and out-migration analyses together suggests that migrants are attracted by market (monetary wage-based) economies, but are unlikely to leave their origin for lack of such; and that employment opportunities are important in both decisions.

The migration rate model presents a similar picture except that variables pertaining to origin cantones are diminished in role. However, this analysis enables comparison with similar applications to Mexico (Greenwood, Ladman, and Siegel 1981) and Venezuela (Levy and Wadycki 1972), thus providing an opportunity to address development aspects of migration at a cross-national level. Although these models differ in variable definition, years studied, number and size of areal units, and length of time over which migration is calibrated, the results are strikingly similar. Migration flows from low to high wage/job opportunity areas, and distance, the statistically most significant variable, is a considerable deterrent to movement. There is substantial difference, however, in levels of explained variance: 0.64 and 0.71 for Venezuela and Mexico, respectively; only 0.33 for Costa Rica. This reflects the conventional model's lack of variables representing non-market aspects of the economy and social factors such as migration chain effects, which are considerably more important in agrarian settings with a lesser level of modern sector economic activity.[17]

The above findings are consistent with the development paradigm of migration and related perspectives in a number of ways. First, differences in forces operating at origin and destination places are directly suggested by such perspectives; for example, the noticeably greater leverage of population pressure

(PRES$_i$) on out-migration. Second, differing degrees of explained variance indicate that the conventional model's market economy orientation is less applicable to Costa Rica than to nations such as Venezuela and Mexico, which are more industrialized and economically integrated, characteristics often seen as indices of development. Accordingly, one may conclude that the role of each migration factor is affected by the development milieu in which it operates; in particular, the degree to which traditional or contemporary elements characterize prevailing economic systems.

Empirical findings: spatially varying parameters of the conventional model

To further explore the apparent links between development characteristics and migration, attention now turns to the conventional model with spatially varying parameters. Implementation employed a stepwise regression procedure with a 0.05 significance test for entering/deleting variables. Separate models were estimated for the out-migration (term (3.5)) and relocation (term (3.6)) dependent variables, using the formulation of (3.3); hence, the parameter of each independent variable could vary from no spatial expression to that of a second order polynomial. The final models (Table 3.2) include those variables significant in the initial conventional model, but only some elements of their potential spatial variation. Discussion primarily revolves around maps of the spatially varying parameters (SVPs), which indicate shifts in the role of each independent variable across the Costa Rican landscape. Occasional attention also is given to SVP maps from related analyses of the same data. The equations from which SVP surfaces were generated are presented in Table 3.3; a base map of Costa Rica (Figure 3.3) will aid the reader in understanding discussion of SVP surfaces.

The SVP model indicates that out-migration is inversely related to the percent of urban-based employment opportunities at the origin (PCUJOB$_i$), directly related to its population pressure (PRES$_i$), and unrelated to average monthly wage per capita (WAG$_i$) (Table 3.2), as in the conventional model (Table 3.1). However, it also shows spatial variation in the role of population pressure (Figure 3.4.A), which is strongly positive in the economic core and adjacent coastal areas but negative in more remote

Table 3.2 Spatially varying parameter models of migration: statistical estimates for Costa Rica, 1968–73

Independent variables[a]	Out-migration analysis				Independent variables[a]	Relocation analysis			
	r	b	Beta	t		r	b	Beta	t
WAG-i	-0.36				DIST-ij	-0.26	-0.76	-0.41	-26.96*
WAGX-i	0.12				DISTX-ij	0.13			
WAGY-i	-0.06				DISTY-ij	-0.03			
WAGXY-i	-0.16				DISTXY-ij	-0.13	-0.01	-0.07	-3.22*
WAGX2-i	0.14				DISTX2-ij	0.00			
WAGY2-i	-0.01				DISTY2-ij	0.00	0.03	0.09	4.63*
PCUJOB-i	-0.53	-0.38	-0.46	-4.05*	POP-j	0.56	0.81	0.45	22.01*
PCUJOBX-i	-0.11				POPX-j	0.11			
PCUJOBY-i	0.07				POPY-j	0.18	0.15	0.14	7.11*
PCUJOBXY-i	0.02				POPXY-j	-0.01	-0.03	-0.17	-5.99*
PCUJOBX2-i	-0.08				POPX2-j	0.24	-0.01	-0.13	-4.37*
PCUJOBY2-i	-0.19				POPY2-j	0.34	0.05	0.16	5.20*
PRES-i	0.43	2.27	0.40	3.07*	WAG-j	0.48	0.71	0.09	2.35*
PRESX-i	-0.11				WAGX-j	0.13	0.26	0.15	4.60*
PRESY-i	0.02				WAGY-j	0.12			
PRESXY-i	0.05				WAGXY-j	0.02	0.12	0.21	4.73*
PRESX2-i	0.02	-0.08	-0.33	-2.66*	WAGX2-j	0.17	0.03	0.14	4.02*
PRESY2-i	0.07				WAGY2-j	0.31			
					PCUJOB-j	0.24	0.20	0.06	2.17*
					PCUJOBX-j	-0.04	-0.28	-0.37	-12.12*
					PCUJOBY-j	-0.01			
					PCUJOBXY-j	0.14	-0.11	-0.24	-6.65*
					PCUJOBX2-j	-0.03			
					PCUJOBY2-j	0.05			
					PRES-j	-0.38	-3.70	-0.16	-7.58*
					PRESX-j	-0.02	-0.67	-0.10	-3.95*
					PRESY-j	-0.19	1.46	0.13	3.83*
					PRESXY-j	-0.12	0.16	0.08	2.29*
					PRESX2-j	-0.14	0.12	0.12	4.76*
					PRESY2-j	-0.29			

$$n = 59 \qquad\qquad\qquad n = 3422$$
$$r^2 = 0.42 \qquad\qquad\qquad r^2 = 0.51$$

Note

(a) See text for definitions of dependent and independent variables. The spatial expansions of each independent variable are as in equations (3.1) through (3.3).

* indicates that variable is significant at the 0.05 level or better.

Table 3.3 Regression-derived equations employed for mapping spatially varying parameters (SVPs)

Equations	Analysis and figure where map appears
Out-migration dependent variable	
SVP for PRES-i $\quad = 2.27 - 0.08X^2_i$	(1)[a] 3.4.A
SVP for PCUJOB-i $= -0.56 + 0.01X^2_i$	(2) 3.4.B
Relocation dependent variable	
SVP for DIST-ij $\quad = -0.76 - 0.01X_jY_j + 0.03Y^2_j$	(1) 3.5.A
SVP for POP-j $\quad = 0.81 + 0.15Y_j \quad - 0.03X_jY_j - 0.01X^2_j + 0.05Y^2_j$	(1) 3.5.B
SVP for WAG-j $\quad = 0.71 + 0.26X_j \quad + 0.12X_jY_j + 0.03X^2_j$	(1) 3.8.A
SVP for WAG-j $\quad = 2.53 + 0.15Y_j \quad - 0.04X_jY_j - 0.01X^2_j$	(2) 3.8.B
SVP for PCUJOB-j $= 0.20 - 0.28X_j \quad - 0.11X_jY_j$	(1) 3.6
SVP for PRES-j $\quad = -3.70 - 0.67X_j \quad + 1.46Y_j \quad + 0.16X_jY_j + 0.12X^2_j$	(1) 3.7.A
SVP for PRES-j $\quad = -2.60 + 0.46Y_j \quad + 0.13X^2_j - 0.33Y^2_j$	(2) 3.7.B

Note

(a) (1) indicates SVP equation is derived from the analyses reported in Table 3.2.
(2) indicates SVP equation is derived from ancillary analyses.

regions of the country. This indicates that employment opportunities, the basis of $PRES_i$, are important where the market economy dominates, but less so where traditional economic elements play a greater role. Further, that population pressure and out-migration are inversely related at the periphery suggests market economy elements may actually increase out-migration where there is little opportunity for economic mobility, perhaps because of their demonstration effect. Alternatively, this inverse relationship also is consistent with Connell, Dasgupta, Laishley, and Lipton's (1976: chs. 1, 9) observation that poorer people often are not financially able to migrate, even though their economic need to do so is great.

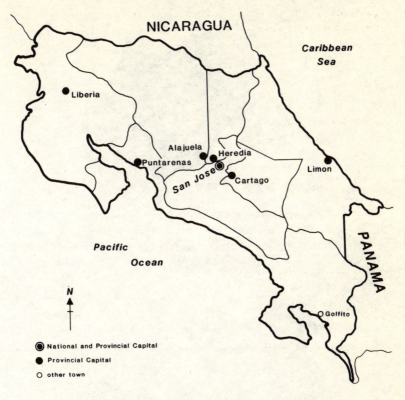

Figure 3.3 Costa Rica base map

Also relevant is spatial variation in PCUJOB$_i$, derived from an SVP out-migration model significant at the 0.10 level (Figure 3.4.B).[18] This pattern is almost identical to PRES$_i$'s (after taking account that signs are reversed throughout the landscape because of PCUJOB$_i$'s opposite relationship with out-migration), and thus reinforces conclusions reported in the preceding paragraph. Broadly speaking, these analyses indicate that conventional models primarily apply to out-migration from core regions where a contemporary market economy prevails, but that a different model and/or set of relationships are relevant for peripheral regions.

Turning now to relocation, the SVP model finds that choice of a destination is inversely related to its population pressure (PRES$_j$) and distance from the origin place (DIST$_{ij}$); directly related to its

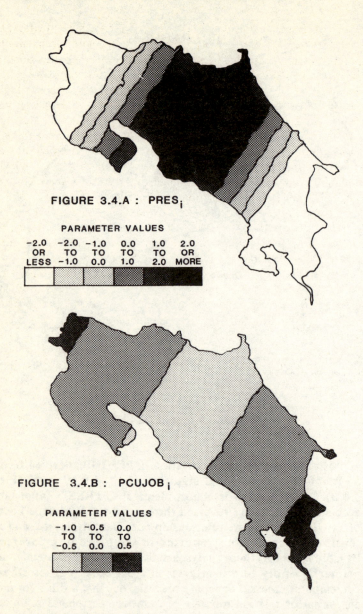

FIGURE 3.4.A : PRES$_i$

PARAMETER VALUES

| -2.0 OR LESS | -2.0 TO -1.0 | -1.0 TO 0.0 | 0.0 TO 1.0 | 1.0 TO 2.0 | 2.0 OR MORE |

FIGURE 3.4.B : PCUJOB$_i$

PARAMETER VALUES

| -1.0 TO -0.5 | -0.5 TO 0.0 | 0.0 TO 0.5 |

Figure 3.4 Spatially varying parameters for out-migration: population pressure and percent of urban-based jobs at the origin

population size (POP_j), wage level (WAG_j), and percent of urban-based jobs ($PCUJOB_j$) (Table 3.2), as in the conventional model. Spatial variation in the role of these variables also is present. Distance, for example, is a fairly uniform deterrent to migration over most of Costa Rica, but noticeably less so in the Caribbean coastal area and in the Golfito region bordering Panama (Figure 3.5.A). Both experienced significant stimuli to economic development in the 1968–73 period: in the Caribbean cantones, extensive expansion of commercial banana cultivation (Blutstein, Andersen, Betters, Dombrowski, and Townsend 1970; Hall 1985: 167–70; West and Augelli 1976: 448–61); and in the Golfito area, completion of a Pan American Highway segment linking the core of Costa Rica with Panama (Blutstein, Andersen, Betters, Dombrowski, and Townsend 1970: 203–4). In both instances, then, the resultant spurt in economic opportunity was sufficient to offset the normal friction of distance as it pertains to migration.

With regard to POP_j (Figure 3.5.B), the most common relationship is in the Meseta Central, Costa Rica's core region with a long tradition of urban-based small-scale commerce and informal activity, for which total population is a surrogate. However, POP_j exerts considerably greater leverage on migration to the Golfito and Caribbean coastal cantones, again reflecting completion of Costa Rica's portion of the Pan American Highway, agricultural expansion, and the resultant increment in economic opportunity. It also is noteworthy that population size relates inversely with migration to cantones bordering Nicaragua, which is largely a cattle producing area characterized, prior to 1968, by traditional latifundia practices. Costa Rican planning agencies sought to improve productivity of this industry in the 1960s, in part through stimulating contemporary agricultural practices, but their actions resulted in curtailing subsistence activities, a consequent reduction in the peasant's resource base, and out-migration (Hall 1985: 183–90, 199–204; Seligson 1980; Taylor 1980).[19] In such a setting, in-migration would be dampened (rather than encouraged) by larger populations for two reasons: first, because of an already limited resource base; second, because of a diminished per capita market potential for informal and small-scale commercial activities.

The percent of urban-based employment ($PCUJOB_j$) shows a pattern of variation that is somewhat similar to POP_j. Its greatest leverage is on migration to Caribbean cantones, while an average relationship is found in the Meseta Central (Figure 3.6). Howev-

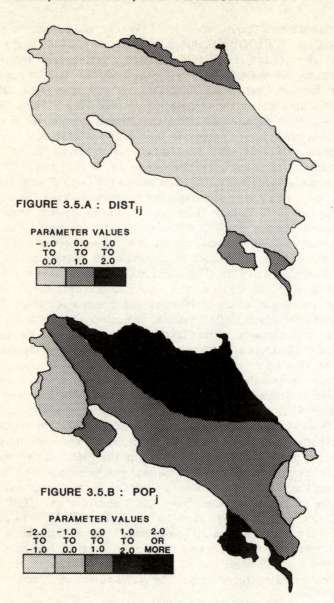

FIGURE 3.5.A : DIST$_{ij}$

PARAMETER VALUES

-1.0 TO 0.0	0.0 TO 1.0	1.0 TO 2.0

FIGURE 3.5.B : POP$_j$

PARAMETER VALUES

-2.0 TO -1.0	-1.0 TO 0.0	0.0 TO 1.0	1.0 TO 2.0	2.0 OR MORE

Figure 3.5 Spatially varying parameters for relocation:
distance between origin and destination and
population at the destination

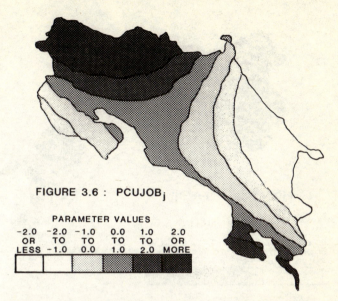

FIGURE 3.6 :   PCUJOB$_j$

PARAMETER VALUES

-2.0 OR LESS	-2.0 TO -1.0	-1.0 TO 0.0	0.0 TO 1.0	1.0 TO 2.0	2.0 OR MORE

Figure 3.6 Spatially varying parameters for relocation: percent of urban-based jobs at the destination

er, a significant inverse relationship appears in the cantones bordering Panama, an area of cacao and banana cultivation in decline from the 1960s onward (Blutstein, Andersen, Betters, Dombrowski, and Townsend 1970; Hall 1985: 170–2; West and Augelli 1976: 448–61); and a direct relationship is found in some cantones bordering Nicaragua, described in the preceding paragraph. Similar levels of PCUJOB$_j$, then, discourage in-migration to remote cantones that are declining, but encourage it to economically ascendant ones. Hence, Costa Rican migrants seem sensitive and responsive to future as well as present opportunity, as posited by the human capital conceptualization discussed earlier in this chapter.

In considering the role of population pressure, two maps are used. One is as above, from the model in which SVPs for all independent variables are derived simultaneously (Figure 3.7.A); a second map is from a model that includes all independent variables but derives SVPs only for PRES$_j$ (Figure 3.7.B).[20] Especially interesting is the direct relationship between PRES$_j$ and migration to peripheral cantones bordering Nicaragua and Panama

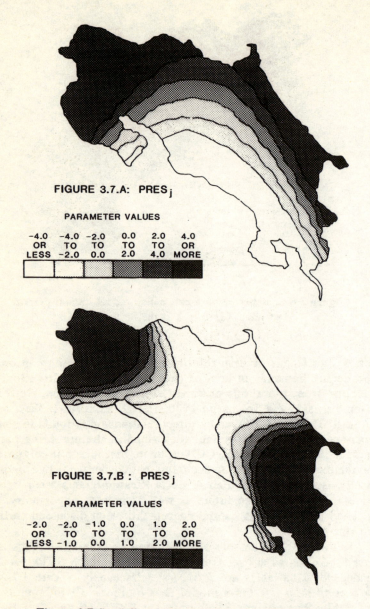

FIGURE 3.7.A: PRES_j

PARAMETER VALUES

-4.0 OR LESS	-4.0 TO -2.0	-2.0 TO 0.0	0.0 TO 2.0	2.0 TO 4.0	4.0 OR MORE

FIGURE 3.7.B : PRES_j

PARAMETER VALUES

-2.0 OR LESS	-2.0 TO -1.0	-1.0 TO 0.0	0.0 TO 1.0	1.0 TO 2.0	2.0 OR MORE

Figure 3.7 Spatially varying parameters for relocation:
population pressure at the destination

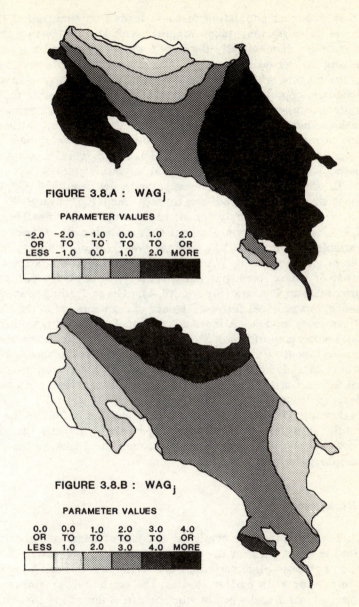

Figure 3.8 Spatially varying parameters for relocation: average
monthly wage per capita at the destination

(that is, higher population pressure leads to higher, rather than lower, in-migration), taken together with knowledge that these areas were economically depressed and had high out-migration during the period in question. These facts suggest a segmented labor market wherein some are in-migrating in response to economic opportunity, while others already in the destination are under- or unemployed, maintaining in a traditional economy fashion, out-migrating, or some combination thereof. Although not surprising, given dual society models of development and empirical research on Third World labor markets, this has not been observed through conventional migration studies.

To illustrate the role of average wage per capita (WAG_j), two maps are again employed: one from the model in which SVPs are derived simultaneously for all independent variables (Figure 3.8.A); a second from a model that includes all independent variables but derives SVPs only for WAG_j (Figure 3.8.B). Under the original treatment, WAG_j directly relates to in-migration over most of Costa Rica, but considerably more so for cantones proximate to Panama (Figure 3.8.A). Given findings reported above, this provides further evidence of segmentation in the Costa Rican labor market. A more conventional picture is provided by the second map of WAG_j SVPs (Figure 3.8.B); this shows wages are an incentive for in-migration primarily to the core region, where market economies dominate, and to the economically ascendant Caribbean and Golfito cantones. Further, both maps derived from analyses with only one set of spatially varying parameters (Figures 3.7.B for $PRES_j$ and 3.8.B for WAG_j) are similar in pattern, and together support the hypothesis that conventional migration factors are most operant in core, rather than periphery, locales.

Discussion

This section presented empirical analyses of intercantonal migrations in Costa Rica for the period 1968 to 1973, focussing on the role of place-to-place variations in development milieu. To enable comparison with earlier studies, the methodology employs a conventional model wherein migration from places i to j is related to their wage rate differential, job opportunity differential, and levels of information concerning economic conditions and related amenities. That model is altered, however, by allowing its b-coefficients of regression to vary spatially, making them a function of X–Y locational coordinates associated with each observation. A second variation is estimating the model for two complementary dependent variables: the probability of out-migration from i, and the probability of relocation to j given that migration has been decided. When combined as joint probabilities, these form the migration rate, the conventional dependent variable, but separate treatment brings conceptual consistency with the two-stage process posited by Brown and Moore (1970) and extended to Third World settings by Brown and Sanders (1981). This approach also highlights origin push factors, which development scenarios see as highly significant, but which migration rate formulations tend to mask.

Four sets of conclusions emerge. First, for both out-migration and relocation there is spatial variation in the role of most variables, including instances where the sign of a variable's coefficient actually reverses from one location to another. Parameter variation also occurred in a reasonably consistent core-to-periphery pattern, as anticipated in the development paradigm of migration discussion above.[21]

Second, in further elaborating the development–migration link, attention must be given to migration-relevant events that occur in less predictable spatial patterns. In this Costa Rican example, such events include growth of commercial banana cultivation in Caribbean cantones, decline of commercial cacao cultivation in cantones proximate to Panama, land reform to improve cattle industry productivity in cantones bordering Nicaragua, and completion of the Pan American Highway link between Costa Rica and Panama. Labor market segmentation and social structure bifurcation also were seen as important. The specific nature and location of these events are not readily predictable, but they represent basic elements of Third World development scenarios: changes in commercial, export-oriented agriculture, infrastructure development, land reform, and economic/social class distinctions that often are prejudicial. Further, such occurrences are likely to impact migration in a different manner than conventional models suggest. For example, in peripheral areas with little economic opportunity, presence of market economy elements, such as urban-based employment, appears to encourage (rather than discourage) out-migration, perhaps because of demonstration effects. Similarly, in peripheral areas experiencing a surge of economic activity, population size, wage levels, and urban-based employment had more than the usual leverage on in-migration, while distance between origin and destination was less of a deterrent. Finally, portions of the population may behave differently under similar conditions due to labor market segmentation.

Third, these analyses demonstrate that ignoring development effects on migration omits a fundamental dimension of the process. Accordingly, conclusions drawn from conventional models, many of which have been utilized in policy formulation, are likely to be misleading because usual procedures embody the urban bias noted by Lipton (1976), Todaro and Stilkind (1981), and many others. In the present case, for example, conformance to conventional findings is most evident in the San Jose/Meseta Central core region, whereas different results obtain for other portions of Costa Rica. More generally, a need to better understand the Third World periphery is once more in evidence.

Finally, models of migration should consist of variables that better specify operant processes; the standard litany of wage rates, employment opportunities, and accessibility is no longer sufficient. More fully elaborating a development paradigm of migration for a spatial frame of reference, and further application of techniques such as those employed here, should aid the task. This theme is explored in the following section and subsequent chapters.

Alternative approaches to studying migration in the aggregate

The above analyses illustrate both the relevance and limitations of conventional approaches to studying migration in the aggregate. Variables such as job opportunities, wage rates, and information flows have noteworthy effects on migration streams. When these represent the range of one's analytical framework, however, other factors are masked; factors that add richness to our understanding of human movements in Third World settings. Examples from the preceding analyses of Costa Rica include changes in export agriculture, infrastructure development, land reform, economic and social class distinctions, labor market segmentation, and the like. Situations to which conventional modeling is less applicable also were identified; for example, rural areas and nations with more agrarian economies.

At one time, demonstrating that a single (conventional) framework applies to many settings was an important task, but that has been accomplished. It is now appropriate to take the next step by incorporating place and process specificity into our knowledge of Third World migrations. Doing so is unlikely to provide the level of generality associated with conventional modeling. Critical, therefore, are procedures, or a methodology, which provide specificity but avoid inundation by idiosyncratic events. An example is provided in the previous section, wherein a conventional model with spatially varying parameters was estimated, followed by identification of place-specific occurrences accounting for shifts in the role of conventional variables.

Another example is provided by Brown and Lawson's (1985b) study of migration directionality in Costa Rica, also for the 1968–73 period. Their first task is to establish the magnitude of rural-to-rural and urban-to-rural migration flows which, if sizable, are counterintuitive to conventional model expectations. In fact, once becoming a migrant, 51.5 percent of Costa Ricans originating from San Jose relocated to rural areas, 39.2 percent of those originating from urban areas that also are province capitals, 38.9 percent from other urban areas, and 40.2 percent from rural areas.[22] Similar findings have been reported by Taylor (1980), also for Costa Rica; Kay (1982) for Chile; Peek (1982) and Preston (1980; Preston and Preston 1983; Preston, Taveras, and Preston 1979) for Ecuador. Fifty percent rural-directed migration is not unusual. Hence, conventional wisdom, suggesting a preponderance of rural-to-urban flows in Third World settings, is not universally applicable. More specifically, significant rural-directed migration may be typical of agrarian Third World economies such as Costa Rica or Ecuador; conventional expectations typical of more industrialized countries such as Brazil, Mexico, or Venezuela; and rural-to-urban transfers of population in both settings may result from factors fomenting rural out-migration, as readily as from destination pulls.

Brown and Lawson's second task is to identify factors underlying rural-directed/originated migrations. To obtain a reasonable level of generality, salient destination and origin cantones were set out from others; then examined qualitatively (through regional geography accounts, thematic maps, topical publications, and the like) for occurrences during the 1960s and early 1970s that account for higher than expected migration levels. In-migration was attributed to three general factors: (1) well-established but expanding commercial agriculture (banana production in Caribbean cantones, following introduction of disease resistant varieties; cattle in the north, stimulated by government policy; and cacao and sugar cane, resulting from price increases in world markets); (2) strength in secondary and tertiary activity of the rural-nonfarm sector (stimulated by location in proximity to San Jose and/or by backward and forward linkages with agriculture and bauxite mining); and (3) transportation (lying astride major road and/or rail arteries, completion of the Pan American Highway link between Costa Rica and Panama as the result of government and donor-nation actions, government development efforts in Caribbean and northern cantones). Out-migration was attributed to (1) land consolidation, (2) curtailment of commercial agricultural activities (banana production in Pacific cantones near Panama), and (3) government-supported expansion of commercial cattle production (which replaced labor-intensive traditional systems and involved land reform that reduced subsistence agriculture opportunities and other means by which peasants supplemented their income).

While these factors in general pertain to economic opportunities, the greater level of specificity *deepens* our understanding of development as it *actually* occurs and migration responses thereto. Especially interesting is the *cyclical*, rather than development, underpinnings of several migration factors; also that these factors reflect *world economic conditions*, *government policy*, and *donor-nation actions*, forces highlighted in earlier chapters.[23]

Finally, Brown and Lawson demonstrate that modeling can be more authentic than under conventional approaches, even though generalization is required, and more effective. Using a formulation identical to that reported in Table 3.1 above, they obtained an r^2 of 0.54 for migration streams from all cantones as origins to urban destinations, but only 0.19 for all origins to rural destinations. Then, to more accurately portray variables pertinent to rural flows, the model was modified by drawing on qualitative analyses summarized in the preceding paragraphs. Remaining from the conventional model were intercantonal distance, canton wage level, and canton population pressure. Added was the magnitude of a canton's rural labor market, measured as the number employed in agriculture and mining; the magnitude of a canton's rural-nonfarm or urban labor market, measured as the number employed in manufacturing, commerce, and service; percent change in a canton's share of the rural labor market from 1963 to 1973; percent change in the rural-nonfarm/urban labor market; change in rural land distribution between 1963 and 1973, measured by gini coefficients; and a migration chain variable, measured by movements before 1963 to avoid overlap with the dependent variable based on 1968–73 migrations.

Application of this model increased explained variance to 54 percent or better. Rural out-migration varied directly with the level of earlier movements, population pressure, and land consolidation, but was deterred by the magnitude of, and positive changes in, rural-nonfarm/urban employment. Rural in-migration varied inversely with intercantonal distance; directly with the level of earlier movements, wages, and the magnitude of, and positive changes in, agricultural labor. Additionally, migration streams from urban origin cantones to rural destinations varied directly with the magnitude of rural-nonfarm/urban employment, while rural origin streams were deterred by high levels of population pressure; an interesting distinction. Finally, migration chain effects, a social component, were at least as strong as economic forces; these ranged from approximately equal importance in out-migration, to somewhat greater importance in urban-to-rural, and were most conspicuous in rural-to-rural migration.

Summary and concluding observations

This chapter examines the effect of place variations in development on *aggregate* migration flows. It first describes the statistical framework most commonly used to study Third World migrations, referred to as conventional modeling. This assumes the same variable set is applicable to all settings, regardless of their development milieu. Second, that development and migration are interrelated is demonstrated by reference to existing literature; also that this linkage is implicit in conventional models. Presented next is a conceptual framework, a development paradigm of migration, which illustrates that the role of conventional model variables shifts under different development milieus. That such shifts actually occur is empirically verified in the fourth section. A regression model with spatially varying parameters, applied to aggregate migration flows in Costa Rica, demonstrates that the roles of conventional variables shift locationally, that these variations may be interpreted in terms of events identified with development, and that development effects of this sort are a fundamental element of migration processes. The final section discusses alternative approaches to studying migration in the aggregate. Needed is a deeper understanding of development as it actually occurs and of migration responses thereto. Achieving such specificity without becoming mired in idiosyncratic events requires a research protocol that raises and addresses questions about broad relationships (rather than taking them as given), an important element of which is a *sense of place* and/or geography procedures for obtaining it.

A research protocol of this type is illustrated by the present chapter. Its application led to the important conclusion that place characteristics identified with development generally result from *government policy, donor-nation actions,* or *cyclical* occurrences such as price shifts in the *world market*; thus supporting the perspective presented in Chapter Two.

On this basis, furthermore, the portrait of development–migration linkages outlined earlier in the chapter can now be embellished. First, internal variations in development generally follow a core–periphery pattern, as already noted. Second, if a country's economic fortunes rise or fall as a result of external cyclical or development forces, the internal pattern need not change; that is, change in each locale may parallel national change. This is unlikely, however, because economic structure differs from place to place. For example, world price increases for sugar cane and cacao brought growth to Costa Rican cantones producing those crops. Similarly, employing a disease resistant banana led to production increases in cantones bordering the Caribbean, decreases in cantones bordering the Pacific, and parallel impacts on the economic fortunes of each area. Third, national policy and donor-nation actions also will affect places differently. Promotion of cattle production in Costa Rica and attendant land reform measures, for example, impacted primarily on cantones in Guanacaste province; while completing the Pan American Highway near Costa Rica's Panama border impacted primarily on cantones in that vicinity and other cantones astride the Highway. Fourth, areas also may grow as a result of local endowments such as a resourceful population. Fifth, development effects on migration (or on other phenomena) must, therefore, be delineated in terms of both general spatial processes and locale-specific characteristics.[24] Methodologies appropriate to this task, if aggregate phenomena are being examined, were discussed in the present chapter. Methodologies for examining individual behavior are put forth in Chapters Four, Five, and Six.

4   Individual migration and place characteristics related to development in Venezuela

Chapter Three demonstrates that place characteristics associated with development affect migration processes in the aggregate. This focus is now broadened by considering development effects on *individual* behavior, as reported in two empirical studies of Venezuela. *Migration* is examined in the present chapter; while Chapter Five considers a set of *labor market experiences* – educational attainment, labor force participation, and wages received. These are, of course, complementary social phenomena. That is, migration is a means of improving educational attainment, employment opportunity, and/or wages; but better labor market experiences, whatever their genesis, also are progenitors of migration.

By examining these phenomena, we see that individuals with identical personal attributes, but residing in different locales, might act dissimilarly because each is faced with different structural conditions, or *development contexts*. Stated simply, it could be said that where one lives affects how one lives. This is particularly evident from Chapter Five's study of labor market experiences, an important component of individual welfare. Further, while the link between place characteristics related to development and personal behavior underlies aggregate migrations examined in Chapter Three, it is brought to the surface by focussing on individuals in Chapters Four and Five.

Another step forward concerns the *measurement* of development. Chapter Three delineated development characteristics on a *qualitative* basis – coupling shifts in the role of conventional migration factors with government-supported growth in commercial cattle production, expanded cacao and sugar cane production resulting from price increases in world markets, land consolidation related to government economic initiatives, completion of the Pan American Highway link between Costa Rica and Panama fomented by government and donor-nation actions, locational shifts in banana production resulting from the introduction of disease resistant varieties, and urbanization. Such occurrences represent the local manifestation of national policies, donor-nation actions, and/or world economic conditions – exogenous forces that are foci of the book's perspective on development.

Having established the appropriateness of this orientation through *qualitative* means, Chapters Four and Five give attention to *quantitative* indices which provide numerical measures for subnational units. Such indices are substantively more general than the development occurrences documented earlier. But as a compensating factor, numerical measures can be used in statistical analyses as variables representing the local articulation of exogenous forces. A more detailed discussion appears later in this chapter.

Development context and individual migrations[1]

Migration, in Third World settings as elsewhere, results when opportunities provided by geographic places are not commensurate with the personal need(s) and/or capabilities of their residents. Research focussing on this interdependency between place and individual attributes is, however, a rare occurrence. More common are formulations derived from economics which examine aggregate migration flows in terms of areal characteristics such as wage rates, employment opportunities, amenity levels, movement costs, and information flows, i.e., the conventional modeling approach described in Chapter Three. Alternatively, formulations derived from demography/sociology often focus on individual propensities to move as a function of personal attributes such as age and educational attainment (Connell, Dasgupta, Laishley, and Lipton 1976; Findley 1977).

Examining place and individual characteristics as interdependent provides a different picture than either of these more common approaches. For example, while educational attainment may be directly related to a person's likelihood of out-migration in all settings, the relationship's strength should vary according to place-specific levels of employment opportunity (Findley 1982). Likewise, although migration streams tend to flow from low to high wage/job opportunity/amenity areas, the applicability of this generalization depends on the educational attainment, age, employment status, and occupational mix of individuals comprising an area's population (Glantz 1973).

The study reported here examines out-migration in Venezuela as a function of both personal attributes and place, or *contextual*, characteristics related to development, thus melding the aggregate and individual perspectives. The basic argument is similar to Chapter Three's. Economic growth involves structural changes in society, which in turn alter the role of specific migration factors. Hence, understanding migration in Third World settings necessarily involves accounting for development effects; and since the character of development varies from place to place, as well as through time, a spatial perspective is mandatory.[2] But whereas aggregate migration flows were considered earlier, the present chapter takes this investigation a step further by examining the manner in which place characteristics related to development impinge, both directly and indirectly through personal attributes, on an individual's decision to migrate. An ancillary issue is the operational definition of development.

Previous research addressing the influence of contextual factors on individual migration in Third World settings is the first topic. Considered next is the issue of development indices and their derivation for Venezuelan political units. Third, out-migration by individuals is related to their age, educational attainment, gender, and development indices portraying place of origin; logistic regression is the analytical tool. The chapter concludes with a discussion of results and contributions to themes of this book.

Background

Three recent studies have been motivated by a research question similar to that addressed in this chapter. These may be summarized in the following manner.

Bilsborrow, McDevitt, Kossoudji, and Fuller (1987) examine the out-migration decision of 3,569 Ecuadorians aged 14 to 27, selected from a sample of 3,427 rural Sierran households.[3] Migrants only included those choosing urban destinations, with 1 indicating such choice and 0 otherwise. Independent variables were the individual's age, education, and marital status; land owned by the farm household; number of adults in the household; and for the canton in which each individual/household resided, distance from Quito (the major destination), absorptive capacity of its agricultural and urban labor market, and the proportion of rural households without electricity as an index of amenity level. Findings concerning contextual or place effects are as follows. Out-migration was less in cantones farther from Quito, in places where a greater proportion of households have electricity, and in cantones with more wage-based agricultural employment. With regard to interactions, an inverse relationship between size of farm and out-migration was characteristic of cantones near Quito, but this declined in strength at greater distances and became reversed in the most remote locales. Other interactions were tested, but found to be not significant.

Findley (1987a, 1987b) examines out-migration among 4,132 individuals in 619 households of Ilocos Norte, a small region in the northern Philippine Islands. The dependent variable is the proportion of family members who migrated or circulated. Independent variables, most of which are calibrated as scales, include: for the family, its class status, human capital attributes, farm enterprise characteristics, previous migration experience, and number of adults; for the community (or context), socioeconomic development level, commercialization of agriculture, economic outlook, physical facilities, accessibility, and intensity of previous migration. Interpretation of statistical results was aided by in-depth family interviews, a particularly interesting and valuable aspect of Findley's work. With regard to contextual effects, out-migration was inversely related to community development level, but directly related to commercialization of agriculture and economic outlook. Noteworthy interactions are as follows. The inverse relationship between socioeconomic development and out-migration was weaker in more accessible locales and in locales with higher levels of previous migration. The positive relationship with commercialization of agriculture was accentuated in communities with more physical facilities. Finally, whereas both upper- and lower-class families were more likely to spawn migrants than those in the middle, this was particularly so in communities with lower levels of previous migration.

Lee (1985) examines the intention to move (or stay) of 1,185 individuals, also from Ilocos Norte in the Philippines. Independent variables indicated a person's commitment to family, job, and place (individual age, gender, marital status, occupation, land ownership, and community links); a person's human capital worth and other attributes facilitating migration (individual educational attainment, previous migration experience, adequacy of economic conditions, and links to places outside the community); and characteristics of the community in which each person resided (its urban status, population density, percent of agriculturalists who are renting, percent of land irrigated, employment and educational opportunities per capita, and indices of electrification, recreation, communication, and transport). Commitment variables were least important; individual resources and community characteristics most important. Among the latter, moving intentions varied directly with urban residence and population density; inversely with the incidence of local employment opportunities and community services. A number of interaction effects were tested, but results were fragmented and difficult to summarize.

Without going into excessive detail, some contrasts between these and the present study should be noted. First, Bilsborrow et al., Findley, and Lee use specially designed surveys; this study is based solely on census data. Second, in part because of data source differences, Findley and Lee are highly localized in orientation; Bilsborrow et al. somewhat so; and all employ household-level variables, particularly Findley and Bilsborrow et al. The work reported here is national in scope and has no household dimension. Third, this study is concerned with individual migrations, treated in dichotomous terms, as is Bilsborrow et al. Findley also is concerned with migration, but focusses on the collective behavior of households, rather than individuals. Lee examines intention to migrate, not migration itself. Fourth, both the present effort and Findley define development as a multifaceted phenomenon, rather than employing single, representative variables. This contributes to achieving generality, which is critical to elaborating a broadly focussed development–migration paradigm. Fifth, these studies, including the present effort, examine similar phenomena from different perspectives and are, therefore, highly complementary to one another. Finally, all four studies provide evidence that contextual or place characteristics associated with development affect the role in migration of individual and household attributes.

Indices of development for sub-national units of Venezuela

While some researchers prefer to depict development through single variables, multivariate scales are favored here. They take into account that aspects of development are highly interrelated, acknowledge its complexity and multidimensionality, and thereby address the *gestalt* or holistic character of development. More central to the present use of multivariate scales, however, is their congruence with this book's perspective on development.

To elaborate, consider procedures originally employed to

delineate modernization surfaces and their change over time. Associated with the modernization paradigm and its view of development as a diffusion process (Brown 1981: ch. 8), these procedures generate *multivariate indices* for *sub-national* areas based on variables pertaining to educational opportunity, economic structure, urbanization, demographic profiles, infrastructures, government institutions, and the like.[4]

Such characteristics are manifestations of, or directly represent, world economic and political conditions, donor-nation actions, and Third World government policies. Further, variables that are surrogates for these broad forces pertain to sub-national units so that the local articulation of exogenous factors is delineated, albeit in a generalized manner.[5] Finally, modernization-type indices measure socioeconomic aspects of places that are clearly relevant to migration and the labor market experiences treated in Chapter Five.

In these studies of Venezuela, then, development impacts are gauged through indices derived from principal components analysis of variables reflecting socioeconomic distinctions among Venezuelan distritos, a local political unit (Table 4.1).[6] Data were provided by published volumes and by individual records of Venezuela's 1971 Census of Population, the latter compiled by Centro Latinoamericano de Demografia (CELADE) and aggregated by distrito.[7]

Considering the variables first, for each distrito they represent its economic structure as indicated by employment in various sectors, educational and occupational attainment, rural—urban character, population mobility (or stability), and other demographic attributes such as its dependency ratio, gender balance, number of children, and population pressure. Variation in these characteristics from distrito to distrito obviously reflects the kinds of exogenous forces that are highlighted in this book.

Two development dimensions, accounting for 66.1 percent of the total variance, were identified. Development dimension one, referred to as STRUCTURE when used as a variable in subsequent analyses of migration and labor market experiences, represents a traditional—contemporary continuum in the socioeconomic structure of Venezuela's distritos. Particularly important in defining this dimension are variables representing economic and educational characteristics of each distrito, which in turn are related to its rural—urban balance.[8] To illustrate STRUCTURE in terms of actual places (Appendix 4.1), high positive component

Table 4.1 Principal components analysis of place variables for 178 Venezuela distritos, varimax rotation[a]

Variable	Mean	Component I	Component II	Commu nality
Percent of population residing in rural areas	39.19	-0.79	0.31	0.72
Percent of population born in same distrito[b]	87.03	-0.74	-0.12	0.57
Percent of population migrants, in residence less than five years[b]	10.06	0.79	-0.11	0.63
Percent of population migrants, in residence more than five years[b]	14.46	0.77	0.07	0.59
Percent of population with no schooling	51.34	-0.84	0.25	0.77
Percent of population college educated	0.54	0.58	-0.37	0.47
Mean years of schooling	2.18	0.86	-0.35	0.87
Percent of employment in primary activity	78.27	-0.89	0.33	0.89
Percent of employment in secondary activity	6.47	0.79	-0.26	0.70
Percent of employment in tertiary activity	15.26	0.81	-0.33	0.77
Ratio of secondary to tertiary employment	0.41	0.39	-0.10	0.16
Mean occupational status (after Treiman (1977), scale from 1 to 100)	31.36	0.82	-0.30	0.76
Mean wage, bolivares per month	511.06	0.85	-0.19	0.75
Mean number of children	2.93	-0.31	0.79	0.72
Population pressure (total population/ employed population)	1.94	-0.08	0.90	0.82
Dependency ratio ((population<15 + population>65)/population aged 15-65)	1.05	-0.34	0.77	0.70
Percent of population female	49.46	0.01	-0.58	0.34
Percent variance explained – by each factor		54.2	11.9	
– cumulative		54.2	66.1	

n = 178

Notes

(a) Distrito-level data not available from published tables was obtained by aggregating individual records, sampled from the 1971 Population Census by CELADE.

(b) These variables reflect two different census questions; one pertaining to place of birth, the other to place of previous residence. Migrants are persons whose previous residence differs from present residence, irrespective of where they were born.

scores (contemporary) are found for distritos representing Venezuela's five largest cities; Caracas (+2.64) in the Distrito Federal, Maracaibo (+1.56) in the state of Zulia, Valencia (+2.36) in Carabobo, Maracay (+2.45) in Aragua, and Barquisimeto (+1.29) in Lara. High negative scores (traditional) are found for distritos with a low percent of urban population; for example, Monagas in Anzoategui (-1.24), Pedro Camejo in Apure (-1.06), and Arismendi in Barinas (-1.13).

Development dimension two, referred to as PRESSURE when used as a variable in subsequent analyses of migration and labor market experiences, represents population pressure within a distrito and/or the degree to which its population is dependent rather than economically active. Particularly important in defining this dimension are variables representing demographic characteristics of each distrito, but these reflect other factors.[9] To illustrate in terms of actual places (Appendix 4.1), high positive principal component scores (high pressure) are found for remote distritos in remote states, generally within Venezuela's southwestern plains region (or Llanos) where, historically, access to major urban centers has been reduced by the intervening Andean Cordillera. Examples include Paez in the state of Apure (+1.70), Romulo Gallegos in Apure (+1.21), Arismendi in Barinas (+1.90), Girardot in Cojedes (+2.27), and Guanarito in Portuguesa (+1.42). High positive scores also are found for economic backwater distritos with ready access to urban agglomerations. Generally, these are located in northeastern states that were major producers of export staples in the early 1900s. Examples include Bruzual and Libertad in Anzoategui (+1.06, +1.05), Anzoategui in Cojedes (+1.86), Bolivar, Caripe, and Cedeno in Monagas (+1.13, +1.15, +1.03), and Andres Eloy Blanco, Cajigal, and Marino in Sucre (+1.63, +1.43, +1.33).

Placing these indices in a broader perspective, the traditional—contemporary continuum represented by STRUCTURE closely corresponds with an area's rural—urban character (as noted for other areas by Berry 1969; Pedersen 1975; Scott 1982; Stohr 1974); while PRESSURE reflects Stohr's (1975; Berry, Conkling, and Ray 1987: ch. 16) classification of locales as being either downward or upward transitional (Figure 2.2). Hence, one might view Venezuelan distritos as falling into four categories based on the values of STRUCTURE and PRESSURE: urban/downward transitional (+/+); urban/upward transitional (+/−), rural/downward transitional (−/+); and rural/upward transitional (−/−).

To visualize this, and indicate the spatial pattern of Venezuelan development, component scores for each distrito were combined (Appendix 4.1) and mapped in contour form (Figure 4.1).[10] Highest levels occur around Venezuela's major urban agglomerations – from west to east, Maracaibo, Barquisimeto, Valencia-Maracay-Caracas, Barcelona-Cumana, Ciudad Guayana-Ciudad Bolivar – and tend to decline monotonically with distance from these areas, conforming with core–periphery expectations (Figure 4.1, inset). Secondary foci of development include San Cristobal and Merida, long-established urban centers in the Andean region; Barinas at the western edge of the Llanos, and Coro near the Caribbean, both associated with Maracaibo's petroleum production complex; and El Tigre-Anaco-Maturin, associated with Barcelona's petroleum production complex and government-supported development of the Guayana region.

Model specification and data characteristics

The initial data on individuals comprised 116,672 records of Venezuela's 1971 Census of Population, which represent all economically active persons in the CELADE sample of 439,815 (see note 7, this chapter). Economically active is defined as persons indicating employment who were fifteen or more years of age at census time, the age criterion used in asking occupation/employment questions.

Only 65,994 of these records were used, under the following rationale. To best interface with indices of distrito development, needed were migrations that preceded the census by only a few years (i.e., recent migrations) and whose origin could be specified at the distrito level. To meet these criteria, data indicate whether respondents had moved more than five years prior to 1971, less than five years, or not at all; their distrito and state of birth; distrito and state of present residence; but only state of previous residence. This last characteristic created a problem if migration was to be analyzed at the more disaggregated distrito level. One solution was to identify migrants by a 'moved within the past five years' response and use distrito of birth to indicate distrito of previous residence. Instead, a more conservative solution was devised: to include such records only if there is correspondence between state of birth and state of previous residence, thus increasing the likelihood that distrito of birth is also the distrito

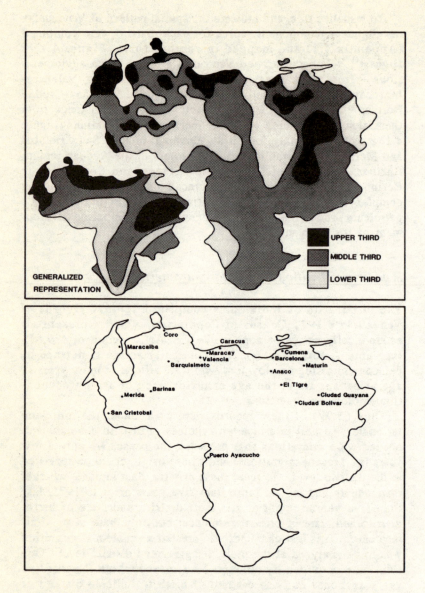

Figure 4.1 Spatial distribution of development in Venezuela 1971,
and selected places

of previous residence. Non-migrants, then, are indicated by records with identical states of present residence, previous residence, and birth. Records not satisfying either (the migrant or non-migrant) criteria were eliminated.

Migrant/non-migrant status ($OUTMIG_i$) is related to an individual's age (AGE_i), educational attainment ($EDUC_i$), gender ($GENDER_i$) and development indices pertaining to his/her distrito of previous residence, i.e., traditional–contemporary aspects of its socioeconomic structure ($STRUCTURE_i$) and its population pressure or degree to which the population is dependent, rather than economically active ($PRESSURE_i$).[11] The analytical procedure is logistic regression, which may be likened to linear regression with a dichotomous dependent variable indicating whether each sampled individual i is an out-migrant (1) or stayer (0).[12] In fact, however, the dependent variable is a ratio of the probability of moving ($Pmov_i$) to the probability of staying ($1 - Pmov_i$), logarithmically transformed; i.e.,

$$\ln(Pmov_i/(1-Pmov_i)) = a + b_1 AGE_i + b_2 EDUC_i + b_3 GENDER_i + b_4 STRUCTURE_i + b_5 PRESSURE_i \quad (4.1)$$

This model was estimated for direct effects only, as above, and for both direct and interaction effects with each place variable. In the latter instance, using equation (4.1) with STRUCTURE as the interaction variable, the model becomes

$$\ln(Pmov_i/(1-Pmov_i)) = a + b_1 AGE_i + b_2 EDUC_i + b_3 GENDER_i + b_4 STRUCTURE_i + b_5 PRESSURE_i + b_6 AGE_i STRUCTURE_i + b_7 EDUC_i STRUCTURE_i + b_8 GENDER_i STRUCTURE_i + b_9 PRESSURE_i STRUCTURE_i \quad (4.2)$$

Zero-order relationships also were estimated; that is, between $\ln(Pmov_i/(1-Pmov_i))$ and each independent variable by itself. Using AGE as an example,

$$\ln(Pmov_i/(1-Pmov_i)) = a + b_1 AGE_i \quad (4.3)$$

These analyses were carried out for the whole sample, for persons from distritos with urban centers of 20,000 or greater population, and for persons from distritos where no urban center exceeds 20,000.[13] This approach provides a general picture of out-migration from Venezuelan distritos and also indicates how

Table 4.2 Mean values of variables related to individual out-migration in Venezuela[a]

Variables	Total sample Non-migr'ts	Migr'ts	Distritos with urban center >20000 popul'tn Non-migr'ts	Migr'ts	Distritos with no urban center >20000 popul'tn Non-migr'ts	Migr'ts
Age (years)	33.63	28.71	32.44	28.54	36.05	28.87
Educational attainment (years attended, 0-16)	3.83	5.19	4.68	5.98	2.10	4.42
Gender (1=male; 2=female)	1.23	1.34	1.28	1.34	1.13	1.34
Traditional-contemporary index (+ = more contemporary distrito; − = more traditional)	0.91	0.41	1.61	1.39	-0.51	-0.54
Population pressure/dependent population index (+ = more pressure in distrito; − = less)	-0.45	-0.28	-0.63	-0.53	-0.07	-0.03
Number of observations	59,425	6,569	39,881	3,244	19,544	3,325

Note

(a) For all migrant–non-migrant comparisons, means are significantly different from one another at the 0.01 level or better.

factors shift importance between a setting of large and intermediate-size urban places and one of small urban places and rural areas, a categorical type of place, or contextual, effect. An initial impression is provided by Table 4.2 which presents mean values of independent variables for migrants and non-migrants in each sample. Migrants are more likely to be female, younger, and better educated. Migrant–non-migrant differentials in gender and age are similar for all samples. In educational attainment, migrants from urban distritos exceed non-migrants by 1.3 years, while the differential is 2.3 years in more rural areas. Concerning contextual variables, migrants tend to originate in places with more traditional socioeconomic structures and higher levels of population pressure, but the differential is more marked in urban distritos. Lastly, these samples indicate the likelihood of out-migration is approximately 0.10 overall, 0.08 for urban distritos, and 0.15 for more rural locales. To embellish this picture, attention turns to findings from logistic regression analyses.

Direct effects

Table 4.3 presents logistic regression results for the total sample and each sub-group, estimating only for direct effects as in Equation (4.1). Log likelihood ratios indicate all models are significant at the 0.01 level or better.

First considering findings for the total sample, coefficients for all variables are significantly different from zero, both in zero-order and multiple variable analyses. As expected, the likelihood of out-migration is greater for females, varies inversely with age, and varies directly with educational attainment, traditionalness, and population pressure. Betas from the multiple variable analysis indicate that traditional–contemporary context is the most important determinant of out-migration, followed sequentially by educational attainment, age, gender, and population pressure. When considered singly, however, the first three of these variables are nearly equal in importance.

Whereas the total sample provides a general picture of out-migration from Venezuelan distritos, sub-sample analyses indicate how factors shift importance in different settings, a categorical type of contextual effect. In this light, consider STRUCTURE. Both sub-sample analyses indicate out-migration probabilities are greater for more traditional (or less contemporary) distritos. However, while STRUCTURE is of primary importance among distritos with urban centers of 20,000 or greater population, elsewhere it operates as a secondary factor. Hence, in rural and less urbanized distritos, where economic opportunity is lacking overall, out-migration is affected relatively little by areal variations in traditional–contemporary aspects of socioeconomic structure, although some differential remains. Among distritos containing intermediate and large urban centers, on the other hand, those which are less contemporary (and primarily, intermediate in size) tend towards out-migration, while more contemporary places (primarily large in size) tend to be destinations in Venezuela's highly focussed economy.

Population pressure also shifts role in different settings. In distritos with no urban center greater than 20,000 population, higher population pressure gives rise to higher levels of out-migration, as expected. In distritos with urban centers greater than 20,000 population, however, PRESSURE shifts sign; i.e., out-migration is more likely from distritos with less population pressure. This occurs because major urban areas have experienced

Table 4.3 Logistic regressions for individual out-migration in Venezuela, direct effects only[a]

Independent variables	Total Sample					Distritos, urban center >20,000 population					Distritos, No urban center >20,000 population				
	Zero order b	r[b]	Multiple variable b	Beta	r[b]	Zero order b	r[b]	Multiple variable b	Beta	r[b]	Zero order b	r[b]	Multiple variable b	Beta	r[b]
Age (years)	-0.02	-26.16	-0.01	-0.56	-20.67	-0.01	-16.45	-0.01	-0.55	-12.98	-0.02	-24.56	-0.01	-0.56	-15.80
Educational attainment (years attended, 0-16)	0.05	27.78	0.07	0.85	35.13	0.05	19.18	0.05	0.74	20.53	0.11	37.35	0.09	0.77	29.26
Gender (1=male; 2=female)	0.27	18.91	0.22	0.31	14.39	0.14	7.12	0.06	0.11	3.12	0.61	29.56	0.42	0.44	18.45
Traditional–contemporary index (+ = more contemporary; – = more traditional)	-0.16	-29.76	-0.27	-1.12	-41.08	-0.15	-14.22	-0.26	-0.86	-17.68	-0.05	-2.91	-0.21	-0.34	-11.15
Population pressure/dependent population index (+ = more pressure; – = less)	0.11	14.64	0.04	0.10	4.11	0.08	6.91	-0.05	-0.14	-2.85	0.03	3.39	0.06	0.14	4.93
n migrants,			6,385	10.0%				3,244	7.5%				3,325	14.5%	
n stayers,			57,656	90.0%				39,881	92.5%				19,544	85.5%	
n total,			64,041					43,125					22,869		
	log likelihood ratio = 3783.4[b]					log likelihood ratio = 1158.3[b]					log likelihood ratio = 2282.2[b]				

Notes

(a) Analyses done with SPSSX Program PROBIT; total sample analyses based on a 97% random sample due to computer system limits on core storage.

(b) Separately from PROBIT, t-values were estimated as the b-coefficient divided by its standard error; Betas were estimated as b multiplied by the ratio of the standard deviations of the independent and dependent variables (S$_x$/S$_y$); log likelihood ratios were computed by the method outlined in Aldrich and Nelson (1984: 55–56). All t-values and log likelihood ratios are significant at the 0.01 level.

excessive in-migration over the years, which led to large percentages of young people and of informal sector employment, factors
that increase population pressure. As economic focal points of
long standing, however, large urban agglomerations continue to
attract a surfeit of migrants.

Educational attainment is the most important variable for
distritos with no urban center greater than 20,000 population, and
second in importance for distritos with urban centers greater than
20,000. Its role is thus consistent across sub-samples, as is that
of AGE. By contrast, GENDER is among the most important
determinants of out-migration in rural and less urban distritos, but
a minor factor elsewhere. This is because females gravitate to,
and tend not to leave, major urban centers where employment
opportunities are more numerous and gender-biased attitudes less
prevalent (Brown and Kodras 1987; Chapter Five of this book).

Interaction effects

The interactions of each variable with STRUCTURE and PRES-
SURE are now considered (Table 4.4). Log likelihood ratios
indicate all interaction models are significant at the 0.01 level or
better.

To interpret interaction effects, the approach of Casetti's (1972,
1982) expansion method is employed. Determining the effect of
age and how it is modified by traditional–contemporary context,
for example, would be done by considering the following term
from (4.2):

$$(b_1 + b_6 STRUCTURE_i) \, AGE_i$$

or more generally,

$$(b_{dir} + b_{int} STRUCTURE_i) \, AGE_i \tag{4.4}$$

where b_{dir} is the *direct* effect coefficient and b_{int} the STRUC-
TURE*AGE *interaction* term coefficient. Further, since STRUC-
TURE values are component scores, a hypothetical contemporary
place with score +1.0 versus a hypothetical traditional place with
score -1.0 can be assumed. Then, term (4.4) becomes either

$$(b_{dir} + b_{int}) \, AGE \tag{4.5}$$

for the contemporary case, or

$$(b_{dir} - b_{int})\ AGE \tag{4.6}$$

for the traditional, with b_{dir} indicating the direct effect and b_{int} indicating how this is modified by either contemporary or traditional context. PRESSURE may be viewed in a similar fashion.[14] The resulting net values for each variable and each analysis, referred to below as *summed b-coefficients*, are shown in Table 4.5.

First considering the total sample, females are markedly more likely to out-migrate from traditional than from contemporary locales, the summed b-coefficients being, respectively, +0.49 (i.e., 0.32+0.17) and +0.15 (i.e., 0.32−0.17); females also are more likely to out-migrate from high population pressure than from low population pressure settings (summed b-coefficients of +0.39 and +0.13, respectively). By contrast, place characteristics have minimal effect on the inverse relationship between age and out-migration; likewise for the direct relationship between educational attainment and out-migration. Also interesting is the finding that in traditional settings higher levels of PRESSURE increase out-migration with a summed b-coefficient of +0.06, but the relationship is reversed in contemporary distritos where the summed b-coefficient is -0.02.

With regard to interaction effects within sub-samples, in both largely urban (distritos with urban centers greater than 20,000) and largely rural (no such centers) locales, AGE shows little variation by place characteristics. Interesting, however, is the behavior of educational attainment. EDUC is directly related to out-migration in all settings, but among largely urban distritos the relationship is stronger in more developed places. Although this is opposite to conventional expectations, it coincides with findings by Brown and Kodras (1987) and Brown and Lawson (1989) that core areas in Venezuela supply quality human resources to distritos in the periphery.

Concerning GENDER, females are considerably more likely to out-migrate from traditional and/or high population pressure settings whether a distrito is urban or rural. In the latter, however, GENDER's influence is noticeably greater. Specifically, its summed b-coefficients in traditional and high population pressure contexts are approximately 0.48 for rural distritos versus 0.24 for urban distritos; while in contemporary and low population pressure

Table 4.4 Logistic regressions for individual out-migration in Venezuela, direct and interaction effects[a]

Independent variables	Total Sample				Distrios, urban center >20,000 population				Distrios, No urban center >20,000 population			
	Trad-contemp		Pop pres		Trad-contemp		Pop pres		Trad-contemp		Pop pres	
	b	[b]	b	[a]	b	[b]	b	[a]	b	[b]	b	[a]
Age (years)	-0.01	-18.45	-0.01	-18.57	-0.02	-10.91	-0.02	-13.83	-0.01	-10.21	-0.01	-15.72
Educational attainment (years attended, 0-16)	0.07	29.74	0.07	29.05	0.04	9.02	0.05	14.38	0.08	20.10	0.09	29.30
Gender (1=male; 2=female)	0.32	16.98	0.26	14.47	0.18	4.78	0.12	4.86	0.36	11.62	0.42	18.44
Traditional-contemporary index (+ = more contemporary; - = more traditional)	-0.09	-3.14	-0.29	-30.97	-0.35	-7.36	-0.30	-17.36	-0.01	-0.09**	-0.20	-10.05
Population pressure/dependent population index (+ = more pressure; - = less)	0.02	2.00*	-0.08	-2.08*	0.06	1.81**	0.18	3.00	0.09	5.28	-0.01	-0.15**
Interaction effects, Traditional-contemporary index and												
Age	0.00	3.58	na	na	0.01	4.77	na	na	0.00	1.21**	na	na
Educational attainment	-0.01	-5.93	na	na	0.01	1.88**	na	na	-0.02	-3.05	na	na
Gender	-0.17	-11.94	na	na	-0.08	-3.55	na	na	-0.14	-3.34	na	na
Population pressure index	-0.04	-6.17	na	na	-0.06	-3.93	na	na	0.05	2.95	na	na
Population pressure index and												
Age	na	na	0.00	-2.72	na	na	-0.01	-6.44	na	na	0.00	0.88**
Educational attainment	na	na	0.00	1.12**	na	na	-0.01	-2.23*	na	na	0.01	2.28*
Gender	na	na	0.13	6.40	na	na	0.10	3.67	na	na	0.03	1.27**
Traditional-contemporary index	na	na	-0.04	-5.58	na	na	-0.06	-3.91	na	na	0.04	2.20*
	LLR=3264.9[b]		LLR=3086.4[b]		LLR=1215.5[b]		LLR=1236.0[b]		LLR=2320.7[b]		LLR=2298.0[b]	

Notes

(a) Analyses done with SPSSX Program PROBIT; total sample analyses based on a 97% random sample due to computer system limits on core storage.

(b) Separately from PROBIT, t-values were estimated as the b-coefficient divided by its standard error, LLRs, or log likelihood ratios, were estimated according to the method outlined in Aldrich and Nelson (1984: 55–56). All log likelihood ratios are significant at the 0.01 level; for t-values, * indicates not significant at 0.01 level, ** indicates not significant at 0.05 level.

Table 4.5 Net, or summed, b-coefficients related to individual out-migration in Venezuela

Variables	Total sample		Distritos with urban center >20000 popul'tn		Distritos with no urban center >20000 popul'tn	
For prototypical traditional and contemporary settings	*Trad*	*Contemp*	*Trad*	*Contemp*	*Trad*	*Contemp*
Age (years)	-0.01	-0.01	-0.03	-0.01	-0.01	-0.01
Educational attainment (years attended, 0-16)	0.08	0.06	0.03	0.05	0.10	0.06
Gender (1=male=1; 2=female)	0.49	0.15	0.26	0.10	0.50	0.22
Population pressure/dependent population index (+ = more pressure in distrito; − = less)	0.06	-0.02	0.12	0.00	0.04	0.14
For prototypical high and low population pressure settings	*High*	*Low*	*High*	*Low*	*High*	*Low*
Age (years)	-0.01	-0.01	-0.03	-0.01	-0.01	-0.01
Educational attainment (years attended, 0-16)	0.07	0.07	0.04	0.06	0.10	0.08
Gender (1=male; 2=female)	0.39	0.13	0.22	0.02	0.45	0.39
Traditional−contemporary index (+ = more contemporary distrito; − = more traditional)	-0.33	-0.25	-0.36	-0.24	-0.16	-0.24

contexts, these coefficients are approximately 0.30 versus 0.06.

Finally, the role of PRESSURE and STRUCTURE in rural distritos should be considered. There, PRESSURE exerts considerably more influence on out-migration in contemporary, rather than traditional, contexts; and STRUCTURE exhibits a similar pattern in low, rather than high, pressure settings. These relationships are opposite to conventional expectations, which would predict higher out-migration from the more disadvantaged places, but consistent with Connell, Dasgupta, Laishley, and Lipton's (1976: 15–18) observation that a lack of knowledge and resources often inhibits out-migration from remote and poorer villages.

Summary and concluding observations

This chapter examines individual out-migration in Venezuela as a function of personal attributes; place, or contextual, characteristics indexing development; and interactions between personal attribute and place variables. Personal attributes, in addition to migrant status, include the age, educational attainment, and gender of 65,994 individuals in 1971. Development indices pertain to distritos, a small political unit. STRUCTURE represents a traditional–contemporary continuum in economic and social structure; a second index, PRESSURE, represents the degree to which distrito population is dependent, rather than economically active, and its level of population pressure. In general, the likelihood of out-migration varied inversely with age, directly with educational attainment, and was greater for females and for persons located in distritos with a more traditional socioeconomic structures and higher levels of population pressure. Five broad observations conclude the chapter.

First, its findings indicate that *place characteristics*, or place differences, associated with development are *exceptionally important* in individual out-migrations. STRUCTURE, along with the personal attribute of educational attainment, had the strongest role when considering direct effects of all variables taken together; and PRESSURE, while not dominant, was nevertheless significant. In analyses focussing on interaction effects, traditional–contemporary aspects of socioeconomic structure noticeably affected the roles of gender and PRESSURE, and population pressure affected the role of gender. Also relevant was the rural–urban nature of distritos, a categorical type of contextual effect. For example, in rural and less urbanized distritos, where economic opportunity is lacking overall, out-migration was only moderately affected by areal variations in STRUCTURE, but this relationship was quite strong in distritos containing intermediate and large urban centers. Similarly, gender effects on out-migration were considerable in rural and less urbanized distritos, but not elsewhere; underlying this are place differences in gender-biased attitudes and employment opportunities for females. That educational attainment emerged as most important among individual attributes also is noteworthy. Such attainment is dependent on an infrastructure providing educational opportunity, and the quality of that infrastructure varies according to the socioeconomic structure of places, as shown in Chapter Five below.

Second, because place characteristics have effects such as those just noted, individuals with identical personal attributes, but residing in different locales, might act dissimilarly in that each would be faced with different structural conditions or development milieus. Alternatively, development context might induce dissimilar individuals to act similarly. The point here is that the *nexus* of personal attributes, place characteristics related to development, and migration behavior is *highly complex*. Clearly, further study is needed.

Third, Chapters Three and Four strongly support the view that understanding migration in Third World settings, both at the aggregate and individual levels, necessarily involves accounting for development effects; and since the character of development varies from place to place, as well as through time, a spatial perspective is mandatory. Together with other research, these chapters also provide considerable information that could be used to sketch a broadly focussed, yet highly detailed *development—migration paradigm* which moves far beyond the discussion in Chapter Three. Although the task is not undertaken here, it is one that should yield significant gains for migration research in Third World settings.

Fourth, a continuing issue is how best to *represent development* for purposes of statistical analysis. This study and Findley (1987a, 1987b) combine variables into *scales* that portray development as a multifaceted, multidimensional phenomenon. Others examining the impact of structural characteristics on individual migrations (Bilsborrow, McDevitt, Kossoudji, and Fuller 1987; Lee 1985) index development by single, representative variables, as do simultaneous equation models concerned with the interplay between migration and development at an aggregate level (e.g., Greenwood 1978). Those favoring single variables consider scales to be arbitrary in definition, amorphous, and inappropriate for evaluating theory or gauging policy impacts. This author disagrees, noting that scales embody a broad range of development characteristics, take into account that aspects of development are highly interrelated, better capture its *gestalt* or holistic character, and are more stable in statistical performance than single variables. In addition, multivariate or scaling approaches have a theoretical justification through the development framework put forth is this book. On balance, then, they are deemed preferable, both for description and as a basis for generalization.

Finally, this chapter provides further evidence in support of the perspective that place characteristics related to *development* are largely an artifact of world economic and political conditions, donor-nation actions, and national policies. In broad terms, these forces have led to urban growth, even when efforts are directed towards agricultural development (Lentnek 1980; Lipton 1976; Taylor 1980; Todaro and Stilkind 1981). Accordingly, Venezuela's development (or modernization) surface (Figure 4.1) peaks at its larger cities, mirroring spatial biases in national policies concerning industrial decentralization, fiscal practices, petroleum activities, and improvement of the Ciudad Guayana and other regions (Brown and Kodras 1987; Brown and Lawson 1989; Jones 1982). This observation at a national scale complements those of Chapter Three concerning local characteristics associated with development in Costa Rica and their effect on migration.

Having detected the imprint of exogenous forces at several spatial scales in a general frame of reference, focus shifts to specific examples of such forces treated in a more detailed manner. That there is spatial differentiation in the local articulation of exogenous forces is demonstrated in this and the preceding chapter, but in a highly generalized fashion that gave little attention to underlying mechanisms. In this regard, Chapter Five is primarily concerned with place effects on individual labor market experiences, using the same development indices as here. Yet it also shows that a major ingredient of the relationship, and migration, was Venezuela's policies to spatially decentralize both educational opportunity and economic activity. Shifting the focus even more, Chapter Six examines Ecuador's land reform policies directly, factors underlying their highly variant local expression, and resultant differences in the dynamics of individual movements from the rural Sierra. Taking yet another step, Chapter Seven considers the relationship between Ecuador's import substitution industrialization policies and regional change.

Appendix 4.1 Development characteristics of Venezuela distritos, 1971

State Distrito	Component score I (CS_1)	score II (CS_2)	Composite development index(CS_1+ (-1*CS_2))	Pct urban	Urban place >20000 population
1 Distrito Federal					
101 Departamento Libertador	2.64	-1.75	4.39	99.96	Caracas
102 Departamento Vargas	1.65	-0.86	2.51	91.67	Caracas
2 Anzoategui					
201 Anaco	1.90	0.72	1.18	86.74	Anaco
202 Aragua	-0.60	0.20	-0.80	43.92	
203 Bolivar	1.34	-0.40	1.74	80.23	Barcelona
204 Bruzual	-0.43	1.06	-1.49	17.01	
205 Cajigal	-1.05	0.99	-2.04	0.00	
206 Freites	0.65	0.11	0.54	55.15	
207 Independencia	0.03	0.79	-0.76	45.49	
208 Libertad	0.42	1.05	-0.63	0.00	
209 Miranda	-0.40	0.79	-1.19	48.31	
210 Monagas	-1.24	1.31	-2.55	0.00	
211 Penalver	-0.03	-0.06	0.03	21.18	
212 Simon Rodriguez	1.69	-0.04	1.73	97.60	Barcelona
213 Sotillo	1.85	0.21	1.64	95.15	Barcelona
3 Apure					
301 Achaguas	-0.95	0.64	-1.59	17.52	
302 Munoz	-1.26	-0.13	-1.13	0.00	
303 Paez	-0.13	1.70	-1.83	22.32	
304 Pedro Camejo	-1.06	0.63	-1.69	0.00	
305 Romulo Gallegos	-0.71	1.21	-1.92	31.09	
306 San Fernando	-0.07	-0.45	0.38	56.37	San Fernando
4 Aragua					
401 Girardot	2.45	-0.72	3.17	98.82	Maracay
402 Marino	1.73	0.52	1.21	90.80	Maracay
403 Ricaurte	1.56	-0.25	1.81	81.78	Maracay
404 San Casimiro	-0.09	-0.07	-0.02	45.69	
405 San Sebastian	0.07	-0.18	0.25	75.27	
406 Sucre	1.57	-0.65	2.22	98.66	Maracay
407 Urdaneta	0.33	0.43	-0.10	38.14	
408 Zamora	0.86	-0.17	1.03	81.59	Villa de Cura
5 Barinas					
501 Arismendi	-1.13	1.90	-3.03	0.00	
502 Barinas	1.07	-0.25	1.32	81.70	Barinas
503 Bolivar	0.03	0.65	-0.62	40.60	
504 Ezequiel Zamora	-0.11	1.20	-1.31	28.53	
505 Obispos	-0.26	0.45	-0.71	24.95	
506 Pedraza	0.16	1.22	-1.06	17.65	
507 Rojas	-0.50	0.63	-1.13	0.00	
508 Sosa	-0.93	-0.62	-0.31	0.00	

Appendix 4.1, continued

State Distrito	Component score I (CS₁)	Component score II (CS₂)	Composite development index(CS₁+ (-1*CS₂))	Pct urban	Urban place >20000 population
6 Bolivar					
601 Caroni	2.55	0.77	1.78	94.08	C.Guayana
602 Cedeno	-1.39	-0.94	-0.45	24.81	
603 Heres	1.20	-0.35	1.55	82.55	C.Bolivar
604 Piar	0.38	0.65	-0.27	54.77	Upata
605 Roscio	-0.25	0.27	-0.52	52.57	
606 Sucre	-0.60	1.31	-1.91	0.00	
7 Carabobo					
701 Bejuma	0.38	-0.45	0.83	55.33	
702 Carlos Arvelo	0.10	0.01	0.09	73.35	
703 Guacara	1.86	0.13	1.73	96.03	Valencia
704 Montalban	0.12	0.12	0.00	73.26	
705 Puerto Cabello	1.83	-0.34	2.17	89.21	P.Cabello
706 Valencia	2.36	-0.96	3.32	95.53	Valencia
8 Cojedes					
801 Anzoategui	0.23	1.86	-1.63	0.00	
802 Falcon	0.52	-0.61	1.13	64.76	
803 Girardot	0.12	2.27	-2.15	0.00	
804 Pao	-0.60	1.43	-2.03	0.00	
805 Ricaurte	-0.87	-0.26	-0.61	0.00	
806 San Carlos	0.32	-0.21	0.53	70.03	San Carlos
807 Tinaco	0.18	0.83	-0.65	51.93	
9 Falcon					
901 Acosta	-0.94	0.24	-1.18	11.69	
902 Bolivar	-0.94	-0.53	-0.41	0.00	
903 Buchivacoa	-1.48	-1.34	-0.14	21.88	
904 Carirubana	1.04	-0.84	1.88	93.81	Punto Fijo
905 Colina	-0.56	-1.21	0.65	59.68	
906 Democracia	-1.21	0.09	-1.30	0.00	
907 Falcon	0.15	-0.97	1.12	30.20	
908 Federacion	-0.87	0.30	-1.17	28.06	
909 Mauroa	-0.92	-1.38	0.46	37.59	
910 Miranda	0.73	-1.00	1.73	87.35	Coro
911 Petit	-1.13	0.44	-1.57	0.00	
912 Silva	0.01	-0.40	0.41	69.86	
913 Zamora	-0.52	-0.49	-0.03	33.88	
10 Guarico					
1001 Infante	0.13	-0.07	0.20	65.60	V.de la Pascua
1002 Mellado	0.64	0.85	-0.21	72.26	
1003 Miranda	0.17	0.31	-0.14	62.67	Calabozo
1004 Monagas	0.06	0.61	-0.55	41.74	
1005 Ribas	-0.66	0.01	-0.67	40.30	

Appendix 4.1, continued

State Distrito	Component score I (CS_1)	score II (CS_2)	Composite development index(CS_1 + ($-1*CS_2$))	Pct urban	Urban place >20000 population
1006 Roscio	0.56	-0.34	0.90	74.11	San Juan de
1007 Zaraza	-0.67	0.06	-0.73	46.61	los Morros
11 Lara					
1101 Crespo	-1.13	-0.63	-0.50	37.31	
1102 Iribarren	1.29	-0.98	2.27	90.36	Barquisimeto
1103 Jimenez	-1.10	-0.62	-0.48	36.06	
1104 Moran	-0.67	-0.57	-0.10	38.77	
1105 Palavecino	0.60	0.06	0.54	62.56	
1106 Torres	-0.20	-0.04	-0.16	37.94	Carora
1107 Urdaneta	-1.65	-0.75	-0.90	18.94	
12 Merida					
1201 Alberto Adriani	1.11	0.35	0.76	65.27	El Vigia
1202 Andres Bello	0.02	1.91	-1.89	0.00	
1203 Arzobispo Chacon	-1.25	0.62	-1.87	0.00	
1204 Campo Elias	-0.87	-1.35	0.48	35.33	
1205 Justo Briceno	-0.81	-0.14	-0.67	0.00	
1206 Libertador	1.65	-2.07	3.72	81.53	Merida
1207 Miranda	-0.35	0.85	-1.20	30.82	
1208 Rangel	-0.92	-0.16	-0.76	0.00	
1209 Rivas Davila	-1.84	-1.84	0.00	0.00	
1210 Sucre	-1.00	-1.04	0.04	14.07	
1211 Tovar	-0.36	-0.40	0.04	38.59	
13 Miranda					
1301 Acevedo	-0.56	-0.15	-0.41	17.55	
1302 Brion	-0.32	-0.52	0.20	48.08	
1303 Guaicaipuro	1.95	-0.62	2.57	81.42	Caracas
1304 Independencia	1.17	0.73	0.44	70.31	Caracas
1305 Lander	0.42	-0.21	0.63	67.54	Caracas
1306 Paez	0.01	0.39	-0.38	37.69	
1307 Paz Castillo	-0.58	-1.30	0.72	45.59	
1308 Plaza	2.16	0.05	2.11	89.88	Caracas
1309 Sucre	3.53	-1.97	5.50	99.32	Caracas
1310 Urdaneta	0.68	-0.89	1.57	53.56	
1311 Zamora	0.95	-0.71	1.66	69.60	Caracas
14 Monagas					
1401 Acosta	-0.72	-0.71	-0.01	28.19	
1402 Bolivar	0.84	1.13	-0.29	67.09	
1403 Caripe	-0.30	1.15	-1.45	18.96	
1404 Cedeno	-0.50	1.03	-1.53	30.57	
1405 Maturin	0.96	0.24	0.72	73.08	C.Guayana
1406 Piar	0.08	0.99	-0.91	20.41	
1407 Sotillo	0.40	0.97	-0.57	43.12	

Appendix 4.1, continued

State Distrito	Component score I (CS_1)	score II (CS_2)	Composite development index$(CS_1+$ $(-1*CS_2))$	Pct urban	Urban place >20000 population
15 Nueva Esparta					
1501 Arismendi	0.55	-1.00	1.55	38.42	
1502 Diaz	-0.23	-0.40	0.17	41.20	
1503 Gomez	-0.16	-1.98	1.82	34.46	
1504 Maneiro	-0.48	-1.46	0.98	44.86	
1505 Marcano	-0.29	-1.61	1.32	59.19	
1506 Marino	0.51	-1.69	2.20	83.94	Porlamar
16 Portuguesa					
1601 Araure	0.27	0.25	0.02	64.77	Araure
1602 Esteller	-0.59	-0.42	-0.17	42.32	
1603 Guanare	0.50	0.34	0.16	53.84	Guanare
1604 Guanarito	-0.33	1.42	-1.75	28.61	
1605 Ospino	-1.19	-0.42	-0.77	25.64	
1606 Paez	1.37	0.11	1.26	85.20	Acarigua
1607 Sucre	-0.78	0.31	-1.09	16.94	
1608 Turen	0.31	1.20	-0.89	49.33	
17 Sucre					
1701 Andres Eloy Blanco	-0.31	1.63	-1.94	47.81	
1702 Arismendi	-1.21	-0.54	-0.67	31.95	
1703 Benitez	-1.34	0.12	-1.46	20.01	
1704 Bermudez	0.07	-0.48	0.55	70.71	Carupano
1705 Cajigal	-1.14	1.43	-2.57	25.22	
1706 Marino	-0.45	1.33	-1.78	25.01	
1707 Mejia	-0.32	0.92	-1.24	33.41	
1708 Montes	-0.65	0.76	-1.41	38.10	
1709 Ribero	-0.89	0.95	-1.84	18.42	
1710 Sucre	0.70	-0.46	1.16	79.30	C.Guayana
1711 Valdez	-0.42	0.81	-1.23	58.06	
18 Tachira					
1801 Ayacucho	0.00	0.17	-0.17	55.17	
1802 Bolivar	0.39	-0.15	0.54	77.54	San Antonio
1803 Capacho	-0.44	-1.24	0.80	38.72	
1804 Cardenas	-0.69	-0.14	-0.55	42.51	
1805 Jauregui	-0.68	-0.10	-0.58	29.17	
1806 Junin	0.15	0.56	-0.41	50.16	
1807 Lobatera	-1.72	-2.21	0.49	0.00	
1808 San Cristobal	1.08	-1.13	2.21	86.32	San Cristobal
1809 Uribante	-1.33	-0.26	-1.07	8.99	
19 Trujillo					
1901 Betijoque	-0.30	0.84	-1.14	36.11	
1902 Bocono	-1.47	-1.41	-0.06	23.30	
1903 Carache	-1.37	-0.35	-1.02	8.77	

Appendix 4.1, continued

State Distrito	Component score I (CS_1)	score II (CS_2)	Composite development index$(CS_1+ (-1*CS_2))$	Pct urban	Urban place >20000 population
1904 Escuque	-1.31	-0.69	-0.62	22.32	
1905 Trujillo	-0.04	-0.61	0.57	53.55	Trujillo
1906 Urdaneta	-1.76	-1.56	-0.20	0.00	
1907 Valera	0.84	-0.63	1.47	84.57	Valera
20 Yaracuy					
2001 Bolivar	-0.38	1.18	-1.56	22.01	
2002 Bruzual	1.23	1.05	0.18	67.36	
2003 Nirgua	-0.65	-0.02	-0.63	41.46	
2004 San Felipe	0.30	-0.88	1.18	70.02	San Felipe
2005 Sucre	-0.91	-0.85	-0.06	43.19	
2006 Urachiche	-0.37	-0.04	-0.33	70.18	
2007 Yaritagua	0.01	-0.13	0.14	71.02	Yaritagua
21 Zulia					
2101 Baralt	0.62	1.25	-0.63	47.49	
2102 Bolivar	1.11	-0.23	1.34	92.29	Maracaibo
2103 Colon	-0.03	0.33	-0.36	44.36	San Carlos
2104 Mara	-0.27	-0.25	-0.02	29.69	
2105 Maracaibo	1.56	-1.27	2.83	98.32	Maracaibo
2106 Miranda	0.16	-0.88	1.04	33.62	
2107 Paez	-0.94	0.04	-0.98	14.91	
2108 Perija	0.47	-0.20	0.67	69.26	Maracaibo
2109 Sucre	0.12	0.80	-0.68	19.77	
2110 Urdaneta	-0.07	-0.96	0.89	65.43	
22 Territorio Federal Amazonas					
2201 Atabapo	-0.76	-1.37	0.61	0.00	
2202 Atures	0.49	0.50	-0.01	65.50	
2203 Casiquiare	-1.18	0.10	-1.28	0.00	
2204 Rio Negro	-1.04	-1.30	0.26	0.00	
23 Territorio Federal Delta Amacuro					
2301 Antonio Diaz	-1.25	0.20	-1.45	0.00	
2302 Pedernales	-0.72	-0.56	-0.16	0.00	
2303 Tucupita	0.77	0.79	-0.02	59.95	Tucupita
24 Dependencias Federales					
2401 Dependencias Federales	1.16	5.96	-4.80	0.00	

MAJOR POLITICAL SUBDIVISIONS OF VENEZUELA

1   DISTRITO FEDERAL
2   ANZOATEGUI
3   APURE
4   ARAGUA
5   BARINAS
6   BOLIVAR
7   CARABOBO
8   COJEDES
9   FALCON
10 GUARICO
11 LARA
12 MERIDA
13 MIRANDA
14 MONAGAS
15 NUEVA ESPARTA
16 PORTUGUESA
17 SUCRE
18 TACHIRA
19 TRUJILLO
20 YARACUY
21 ZULIA
22 TERRITORIO FEDERAL AMAZONAS
23 TERRITORIO FEDERAL DELTA AMACURO

●   STATE CAPITAL
◎   NATIONAL CAPITAL - CARACAS

5 Individual labor market experiences and place characteristics related to development in Venezuela

An important objective of Third World development is improvement in individual welfare. Associated with this are factors such as increased and better quality health care, housing, education, and employment, as well as attitude shifts that encourage achievement and bettering life's circumstances. But development spreads unevenly, altering the character of some locales more than others and creating spatial variation in personal well-being. Moreover, development effects often differ among population segments defined on the basis of gender, age, educational attainment, socioeconomic status, place of residence (e.g., rural/urban), and the like; segments which are referred to more generally as *social categories*.

The effects of spatially uneven development and social category differentials on individual welfare are highlighted in the study reported here. For a national sample of Venezuelans, it examines educational attainment, labor force participation, and wages received; a linked set of attributes which, when considered together, indicate one's experience as a work force member. These are documented at the individual level and coupled with other personal attributes and with place characteristics related to development. More precisely, this study is framed by the following questions. How does location, defined in terms of place characteristics related to development, affect individual labor market experiences? How does social category, taking *gender* as an example, affect labor market experiences?[1] How do earlier labor market experiences affect later ones (for example, educational attainment effects on labor force participation and wages received)?

Concerning themes more central to the book, this chapter contributes in two ways. First, it continues Chapter Four's exploration of development impacts at the individual level, but introduces an *equity* issue. In particular, place characteristics associated with development take on greater relevance because they, as well as social category, affect one's access to opportunities for educational attainment, labor force participation, and wages received. Personal well-being is altered accordingly.

This chapter's second thematic contribution concerns the premise that place characteristics related to development largely represent the local articulation of world economic and political conditions, donor-nation actions, and national policies. Germane to labor market experiences in Venezuela are its policies to spatially decentralize both educational opportunity and economic activity, which were initiated during the 1950s as part of a broad effort to disseminate petroleum wealth benefits (Blutstein, Edwards, Johnston, McMorris, and Rudolph 1977: 84–90, 152–7; Ewell 1984: chs. 4 and 5). While educational opportunity was widely dispersed, raising educational attainment throughout the country, industrial decentralization was limited primarily to the Ciudad Guayana complex and urban areas in proximity to Caracas, the major city (Figure 4.1) (Gwynne 1986: 123–7).

That difference in the scale of decentralization aggravated already existing spatial disparities between employment opportunities and the educated work force, thus increasing structural tensions (Lawson and Brown 1987) and the propensity for urban-directed migration, particularly among females.[2] Hence, specific policies affected the spatial distribution of opportunities for education and employment, of population characteristics (e.g., percent of distrito population that is female), and of population overall – all of which constitute place characteristics associated with development. These policies also underlie the migration findings reported in Chapter Four.

Labor market experiences, location, and social category[3]

Indices of place, or contextual, characteristics pertaining to Venezuelan development were reported in the preceding chapter (pages 83–8; Appendix 4.1). Considered here is their influence on educational attainment, labor force participation, and wages received as documented in census statistics. This is initially viewed in the aggregate by relating average labor market experiences of the distrito to its composite development index (Appendix 4.1). Focus then shifts from a space-economy perspective to consider individual labor market experiences, which are regressed against both personal attributes and development characteristics (indices) of the distrito(s) in which each person resides(ed).

Before reporting these analyses, two observations on earlier studies are appropriate. First, a prominent research concern has been human resources in Third World settings; specifically, the role of education and subsequent labor market experiences in improving a population's skills, abilities, motivations, and ultimately, contribution(s) to economic growth (Brown and Lawson 1989; Fields 1980; Fields and Schultz 1980; Todaro 1985: ch. 11). But the present study highlights a dimension that has received little attention: the role of *place* in human resource formation at the individual level.

Another research genre is concerned with gender differentials in labor market experiences.[4] It would be appropriate to meld this with human resource research by outlining a scenario of labor market experiences in Third World settings that considers both place characteristics and the social category gender, but there are several gaps in previous findings. For example, gender differentials are amply evidenced in terms of a rural–urban dichotomy, but not in terms of other place characteristics (e.g., fine-grained indices of development). Similarly, where spatial regularities in development are interrupted by locationally sporadic occurrences such as mineral exploitation or government projects, wages and labor force participation have been shown to be higher than otherwise expected, but evidence on educational attainment is lacking. Another hiatus concerns the interrelationships between personal attributes and development characteristics, as they affect labor market experiences, i.e., interaction effects. Such gaps are lessened by the research reported here.

Average labor market experiences and Venezuela's space-economy

Three labor market experiences focus this study – individual educational attainment ($EDUC_i$), measured as years of schooling; labor force participation (LFP_i), measured as 1 if in the labor force, 0 otherwise; and wages (bolivares) received per month ($WAGE_i$).[5] The sample for educational attainment and labor force participation consists of CELADE records for all persons aged twenty-one or older at the time of Venezuela's 1971 Population Census; these number 167,949.[6] Analyses of the labor force alone employ a smaller number of records, 56,126. This winnowing reflects both the labor force participation rate of 52 percent (in the sample) and the wage data's incidence of 'no response'.[7]

Appendix 5.1 indicates the mean values of these and other variables used in the study.

To gain an initial indication of development influences on each dependent variable, Figure 5.1 graphs *average* educational attainment, labor force participation, and wages received for women and men in each distrito against its composite development index (see note 10, Chapter Four for index details), using standard ellipses (Brown and Holmes 1971a, 1971b) as a summary measure. Educational attainment, which tends to be slightly lower for females, varies directly with development for both genders. The relationship between labor force participation and development is weak and slightly negative for men, but strongly positive for women. By contrast, male average wages are highly correlated with distrito development; female wages are not. Finally, labor force participation is much lower for females, but wages only slightly so.

The spatial expression of these relationships is indicated by comparing Figures 4.1 and 5.2 which map, respectively, development and (average) labor market experiences for distritos. The pattern of educational attainment for both genders (Figure 5.2.A) closely approximates that of development, although two exceptions are evident. High levels of attainment are found in the vicinity of Puerto Ayacucho, gateway to the Amazon region, and in a wide band covering most of eastern Venezuela; areas targeted by regional development efforts such as the well-known Guayana project (Friedmann 1966; MacDonald 1980; Rodwin 1969). With regard to labor force participation (Figure 5.2.B), its pattern for women closely approximates that of development, whereas the male pattern does not. Standing out in this regard are lesser developed distritos of the Llanos and Amazon regions, reflecting male domination in agriculture, petroleum production, and construction. Wages received (Figure 5.2.C) provide an opposite finding. The spatial distribution of male wages closely approximates that of development; female wages do not.

These findings provoke three general observations, which are addressed more precisely and elaborated in subsequent statistical analyses of individual labor market experiences.

First, earlier research does not provide a basis for hypothesizing educational attainment levels in the vicinity of locationally sporadic development. In Venezuela, where such development has been strongly influenced by regionally-focussed government efforts, the relationship seems distinctly positive. On the other

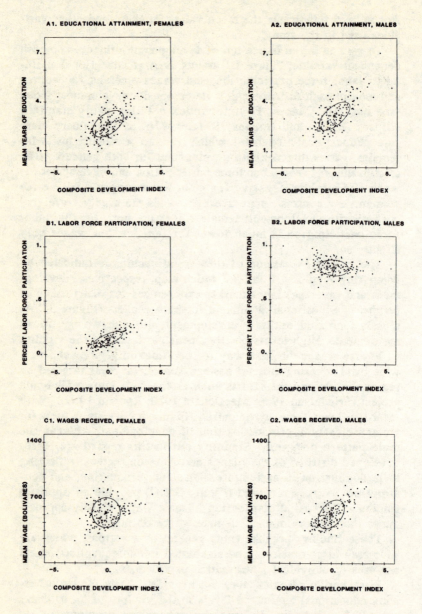

Figure 5.1 Labor market experiences and development in Venezuela, 1971

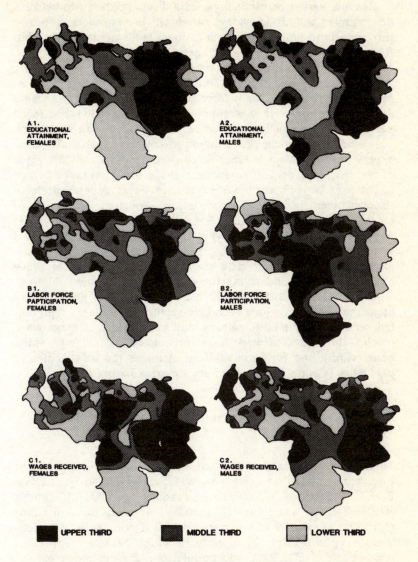

Figure 5.2 The spatial distribution of labor market experiences
in Venezuela, 1971

hand, what if private efforts had more of a role?.

Second, earlier research suggests a direct relationship between development and all dependent variables. In Venezuela, however, this appears to hold only for educational attainment, female labor force participation, and male wages.

The third observation draws on knowledge of Venezuela's space-economy and several findings reported in this section, particularly that male labor force participation and female wages do not correlate with development level and that labor force participation is considerably lower for females than for males. This evidence suggests that a major *crossroad* in the labor market experience of women is labor force participation; and that place of residence is a critical component of the direction taken. Once in the paid labor force, however, female wages appear relatively unconstrained by place characteristics associated with development. For men, on the other hand, labor force participation is widespread, although somewhat restricted in more developed settings where labor supply may be excessive relative to employment opportunities, but male wage levels are affected by place characteristics.[8] Underlying the role of place in these relationships are distrito attributes such as the nature of employment opportunities (e.g., which economic sectors are strongly represented, whether they tend to employ males or females, average wage levels), the degree of work force institutionalization (e.g., union penetration), and for females, local attitudes towards women in paid work (Centro Economico Para America Latina 1983: 180–8).

Research design focussing on individual labor market experiences

To further understand individual educational attainment, labor force participation, and wages received, regression analyses relate $EDUC_i$, LFP_i, and $WAGE_i$ to a person's age (AGE_i); gender ($GENDER_i$), measured as 1 if female, 0 if male; migrant status ($MIGST_i$), measured as 1 if a migrant, 0 if not; traditional–contemporary index for the distrito in which person i resides or resided ($STRUCTURE_i$); and population pressure index for the distrito in which person i resides or resided ($PRESSURE_i$), where distrito indices are identical to those derived in Chapter Four. The result is three models, each representing a *sequential stage* of an individual's labor market experience.

First accounting for educational attainment,

$$EDUC_i = f(AGE_i, GENDER_i, STRUCTURE_i, PRESSURE_i) \quad (5.1)$$

where the distrito variables STRUCTURE and PRESSURE refer to individual i's *previous* residence. Second, labor force participation is examined by

$$LFP_i = f(EDUC_i, AGE_i, GENDER_i, \\ STRUCTURE_i, PRESSURE_i) \quad (5.2)$$

where STRUCTURE and PRESSURE refer to individual i's *present* residence. Third, to account for wage-level differences,

$$WAGE_i = f(EDUC_i, AGE_i, GENDER_i, MIGST_i, \\ STRUCTURE_i, PRESSURE_i) \quad (5.3)$$

where STRUCTURE and PRESSURE refer to individual i's *present* residence.

These equations are estimated by multiple regression techniques for direct effects only, as above, and for both direct and interaction effects with each place variable. In the latter instance, using equation (5.1) with STRUCTURE as the interaction variable, the model becomes

$$EDUC_i = f(AGE_i, GENDER_i, STRUCTURE_i, PRESSURE_i, \\ AGE_iSTRUCTURE_i, GENDER_iSTRUCTURE_i, \\ PRESSURE_iSTRUCTURE_i) \quad (5.4)$$

As in Chapter Four, interaction effects are interpreted with the assistance of Casetti's (1972, 1982) expansion method. To determine the effect of age and how it is modified by traditional–contemporary place characteristics, for example, the following term from (5.4) would be considered:

$$(b_{dir} + b_{int}STRUCTURE_i) AGE_i \quad (5.5)$$

where b_{dir} (or b_1 in (5.4) were it fully specified) is the *direct* effect coefficient and b_{int} (or b_5 in (5.4) were it fully specified) the STRUCTURE*AGE *interaction* term coefficient. Further, since STRUCTURE values are component scores, a hypothetical contemporary place with score +1.0 versus a hypothetical traditional place with score -1.0 can be assumed. Then, term (5.5) becomes either

$$(b_{dir} + b_{int}) \text{ AGE} \qquad (5.6)$$

for the contemporary case, or

$$(b_{dir} - b_{int}) \text{ AGE} \qquad (5.7)$$

for the traditional, with b_{dir} indicating the direct effect and b_{int} indicating how this is modified by either a traditional or contemporary setting. PRESSURE may be treated similarly. The resulting net values for each variable are again referred to as *summed b-coefficients*.

Several characteristics of these equations are noteworthy. First, they represent a sequence of events, both in time and in labor market experience. Education occurs primarily during one's youth, and pertains to a large segment of the population. After completing education, some participate in labor force segments external to the household; others do not. And those entering the labor force receive wages of differing magnitudes. Second, the equations recognize that events are contingent on one another. For example, early educational experience affects both labor force participation and wage level, and earning a wage is dependent on first entering the labor force. Third, events are affected by both personal and place attributes. The latter reflect the spread of development across Venezuela's landscape. Among personal attributes, gender is employed to represent social category effects. Fourth, educational attainment is affected by place attributes where an individual grew up, represented by distrito of previous residence, while labor force participation and wages received are affected by the place he/she lives as an adult, represented by distrito of present residence. Finally, migration is recognized as a mechanism for improving labor market experiences by changing place of residence. Analyses elaborating these and other relationships are now reported.[9]

Statistical analyses: educational attainment

Using the total sample of persons aged twenty-one or more,

educational attainment models were estimated for direct effects only (as in equation (5.1)), and for direct effects and interactions with both the traditional–contemporary and dependent population indices. Results are presented in Table 5.1. F-ratios indicate models are significant at the 0.01 level or better; levels of explained variance are somewhat greater than 19 percent.

Considering the analysis for direct effects only, all variables are significant. Most important are AGE, which varies inversely with educational attainment, and STRUCTURE, which varies directly. These relationships reflect an increase in the availability of education over time, a parallel shift in attitudes towards acquiring formal training, and greater opportunity at all times in more contemporary and urban settings (Blutstein, Edwards, Johnston, McMorris, and Rudolph 1977; Segnini 1974; World Bank 1980).

Concerning interaction effects, place characteristics somewhat alter the role of AGE. With STRUCTURE, its summed b-coefficients are -0.08 (-0.07+(-0.01)) for a prototypical contemporary setting and -0.06 (-0.07+0.01) for a prototypical traditional setting; coefficients for prototypical low and high population pressure distritos are -0.09 and -0.07, respectively. This indicates the spatial pattern of educational attainment (i.e., the level of attainment in one place relative to others) has changed relatively little over time, although more developed settings have slightly improved their position.

Now consider GENDER. Women tend to be less educated than men overall (see direct effect statistics, Table 5.1), but the differential is *notably narrower* in traditional and high population pressure settings (summed b-coefficients of -0.48 and -0.41, respectively) than in contemporary and low population pressure settings (summed b-coefficients of -0.60 and -0.65, respectively).

Taken together, these findings indicate that increased access to education, an accoutrement of development, was distributed uniformly among both less and more developed settings, so that access in the latter remained superior. Nevertheless, in less developed settings female educational attainment improved relative to that of males. Pertinent to these occurrences are Venezuela's efforts to decentralize educational opportunity and access thereto. The outcome seems mixed, leaving spatial disparities intact but impacting females in a positive manner. This conclusion supports both Ewell (1984), who indicates redistribution effects were minimal, and Jones (1982), who suggests the opposite using as an

Table 5.1 Regression statistics accounting for individual educational attainment in Venezuela[a]

Independent variables	Direct effects only				Direct and interaction effects							
					With traditional-contemporary index				With population pressure index			
	r	b	Beta	t	r	b	Beta	t	r	b	Beta	t
Age (years)	-0.34	-0.08	-0.32	-138.35**	-0.34	-0.07	-0.30	-118.94**	-0.34	-0.08	-0.30	-117.15**
Gender (0=male; 1=female)	-0.08	-0.58	-0.08	-34.28**	-0.08	-0.54	-0.07	-28.85**	-0.08	-0.53	-0.07	-27.23**
Traditional–contemporary index, previous residence (+ = more contemporary; – = more traditional	0.29	0.67	0.22	85.77**	0.29	0.98	0.33	44.00**	0.29	0.58	0.19	54.03**
Population pressure/dependent population index, previous residence (+ = more pressure; – = less)	-0.19	-0.34	-0.08	-30.07**	-0.19	-0.29	-0.07	-23.79**	-0.19	-0.58	-0.13	-18.65**
Interaction effects, Traditional–contemporary index and												
Age	na	na	na	na	0.20	-0.01	-0.13	-20.05**	na	na	na	na
Gender	na	na	na	na	0.16	-0.06	-0.02	-4.58**	na	na	na	na
Population pressure index	na	na	na	na	-0.26	-0.10	-0.04	-11.54**	na	na	na	na
Population pressure index and												
Age	na	na	na	na	na	na	na	na	-0.10	0.01	0.05	8.29**
Gender	na	na	na	na	na	na	na	na	-0.09	0.12	0.02	5.80**
Traditional–contemporary index	na	na	na	na	na	na	na	na	-0.26	-0.10	-0.04	-11.56**

Note

(a) Dependent variable is years of education, ranging from 0 to 16. In all regressions, n = 156,924 after eliminating cases with missing data. For each regression, in the order presented from left to right, r²s are 0.19, 0.20, 0.19; Fs are 9,344.88, 5,442.33, 5,385.03.

Significance levels: ** = 0.01, * = 0.05; na = not applicable.

index expenditures on educational buildings which were proportionately greater in low-income regions.

Statistical analyses: labor force participation

Using the total sample of persons aged twenty-one or more, labor force participation models were estimated for direct effects only (as in equation (5.2)), and for direct effects and interactions with the traditional–contemporary and dependent population indices. Results are presented in Table 5.2. F-ratios indicate models are significant at the 0.01 level or better; levels of explained variance are approximately 37 percent.

When examined only for direct effects, all variables are significant. GENDER is most important, coinciding with the earlier observation that labor force participation is less likely for women than for men. Next, but far less significant, participation varies directly with educational attainment and inversely with age.

With regard to interaction effects, traditional–contemporary place characteristics primarily affect the role of GENDER, which shows summed b-coefficients of -0.57 (-0.64+0.07) for a prototypical contemporary setting and -0.71 (-0.64−0.07) for a prototypical traditional setting. The greatest effect of PRESSURE also is on GENDER, yielding summed b-coefficients of -0.53 for a prototypical distrito with low population pressure and -0.71 for a high-pressure setting. Thus, labor force participation by women is more likely in distritos that are more contemporary or have less population pressure.

To further explore gender differences in labor force participation, models also were estimated for males and females separately.[10] That these differ considerably in level of explained variance is itself meaningful. The near zero level for males (r^2=0.03) indicates their labor force participation is widespread geographically, as noted earlier; but it also indicates this occurs irrespective of educational attainment, age, and place characteristics associated with development! Female participation, with an r^2 of 0.14, is constrained by these factors. In particular, educational attainment is most important for women, whereas for men, it is not even statistically significant (at the 0.05 level).

Because the role of place in individual behavior has been an important theme of this book, it is especially interesting to find that residence in a more developed distrito, whether by accident

Table 5.2 Regression statistics accounting for individual labor force participation in Venezuela[a]

Independent variables	Direct effects only				Direct and interaction effects							
					With traditional-contemporary index				With population pressure index			
	r	b	Beta	t	r	b	Beta	t	r	b	Beta	t
Educational attainment (years attended, 0-16)	0.21	0.01	0.10	44.02**	0.21	0.02	0.12	32.14**	0.21	0.02	0.12	37.54**
Age (years)	-0.16	0.00	-0.11	-50.57**	-0.16	0.00	-0.09	-33.12**	-0.16	0.00	-0.09	-35.01**
Gender (0=male; 1=female)	-0.57	-0.56	-0.56	-278.94**	-0.57	-0.64	-0.64	-240.63**	-0.57	-0.62	-0.62	-250.76**
Traditional-contemporary index, present residence (+ = more contemporary; − = more traditional	0.11	0.02	0.06	21.55**	0.11	0.00	0.00	0.52	0.11	0.01	0.02	6.57**
Population pressure/dependent population index, present residence (+ = more pressure; − = less)	-0.07	-0.01	-0.02	-9.43**	-0.07	0.00	0.00	-0.14	-0.07	0.00	0.00	0.33
Interaction effects, Traditional-contemporary index and												
Educational attainment	na	na	na	na	0.18	0.00	-0.02	-3.71**	na	na	na	na
Age	na	na	na	na	0.04	0.00	-0.07	-11.82**	na	na	na	na
Gender	na	na	na	na	-0.18	0.07	0.16	45.58**	na	na	na	na
Population pressure index	na	na	na	na	-0.10	-0.01	-0.05	-13.25**	na	na	na	na
Population pressure index and												
Educational attainment	na	na	na	na	na	na	na	na	-0.15	0.00	0.03	7.58**
Age	na	na	na	na	na	na	na	na	-0.01	0.00	0.07	11.97**
Gender	na	na	na	na	na	na	na	na	0.16	-0.09	-0.13	-40.14**
Traditional-contemporary index	na	na	na	na	na	na	na	na	-0.10	-0.01	-0.06	-14.38**

Note

(a) Dependent variable is 1 for labor force participation, 0 otherwise. In all regressions, n = 156,924 after eliminating cases with missing data. For each regression, in the order presented from left to right, r²s are 0.37, 0.38, 0.38; Fs are 18,225.31, 10,558.24, 10,486.08.

Significance levels: ** = 0.01, * = 0.05; na = not applicable.

of birth or migration, plays a greater role in the labor force participation of women than of men. Underlying this difference is the spatial distribution of employment opportunities for each gender which is indicated by considering, for men and women separately, percent of the labor force employed in each economic sector (Table 5.3). Male employment is disproportionately concentrated in the spatially dispersed activities of agriculture, mineral extraction, construction, and transportation; whereas female employment is disproportionately concentrated in service activities located largely in urban, more contemporary distritos. Similar employment profiles have been found for other nations. Accordingly, the conclusion that residence in a more developed distrito is considerably more important to the labor market experiences of women, than of men, has *wide* applicability.[11]

Another general observation (prompted by the minimal role of AGE and terms interacting it with place variables) is that the historically established spatial form of employment opportunities had changed little by 1971, even though redistribution programs were undertaken more than ten years earlier; that is, 1971 Venezuela had not experienced polarization reversal (Brown and Lawson 1989). Meanwhile, educational attainment increased everywhere, especially that of females in less developed areas. The joint effect of these policy-related occurrences was to exacerbate already existing spatial disparities between employment opportunities and the educated work force, thus increasing migration propensities. This particularly affected women because their employment opportunities were spatially concentrated so that locational change was integral to participation in the labor force.

Statistical analyses: wages received

Using the *labor force* sample only, wage-level models were estimated for direct effects (as in equation (5.3)), and for direct effects and interactions with the traditional–contemporary and dependent population indices. Results are presented in Table 5.4. F-ratios indicate models are significant at the 0.01 level or better. Levels of explained variance are approximately 32 percent.

When examined only for direct effects, all variables are significant. Most important is educational attainment, which varies directly with wages received. At a much lesser but nevertheless notable level of significance, wages are greater for men

Table 5.3 Average monthly wages (bolivares) among branches of economic activity and employment patterns for females and males, Venezuela

Branch of economic activity	Wages	Employment Pattern			
		Females		Males	
Agriculture and fishing	351		3.2%		30.2%
Mining and mineral extraction	1446		0.4%		1.8%
Manufacturing	759				
Foods, drinks, tobacco	725	2.3%		2.6%	
Textiles	625	8.4%		2.8%	
Wood, paper	709	0.8%		2.2%	
Chemicals	908	1.1%		1.1%	
Metal and machinery production	842	1.2%		6.5%	
Other manufacturing	836	0.7%	14.5%	1.1%	16.3%
Utilities	920		0.6%		1.6%
Construction	742		0.7%		7.9%
Commerce	802		9.9%		14.2%
Transportation and communication	859		1.6%		5.9%
Finance, insurance and real estate	1087		1.9%		1.7%
Services	801				
Public administration and defense	858	9.8%		10.6%	
Health and community services	1124	23.4%		5.0%	
Entertainment	943	0.6%		0.8%	
Domestic and personal services	424	33.3%		3.9%	
International organizations	1345	0.1%	67.2%	0.1%	20.4%
Total			100.0%		100.0%

than for women and vary directly with AGE and location in more contemporary distritos.

Concerning interaction effects, returns to educational attainment and age are considerably greater in more developed settings. For AGE, summed b-coefficients are 7.11 (4.08+3.03) and 7.66 (4.81−(−2.85)) for, respectively, prototypical contemporary and low population pressure distritos; versus 1.05 and 1.96 for, respectively, prototypical traditional and high-pressure distritos. For EDUC, parallel summed b-coefficients are 82.91 and 87.12 versus 63.41 and 70.32.

STRUCTURE and PRESSURE effects on GENDER, on the other hand, are counterintuitive in that summed b-coefficients indicate female wage shortfalls are less severe in situations generally seen as more, rather than less, disadvantageous. The seeming contradiction occurs in part because EDUC and AGE covary with STRUCTURE/PRESSURE, but the former two variables

Table 5.4 Regression statistics accounting for individual wages in Venezuela[a]

| Independent variables | Direct effects only | | | | Direct and interaction effects | | | | | | | |
| | | | | | With traditional-contemporary index | | | | With population pressure index | | | |
	r	b	Beta	t	r	b	Beta	t	r	b	Beta	t
Educational attainment (years attended, 0-16)	0.52	85.22	0.54	126.05**	0.52	73.16	0.47	77.66**	0.52	78.72	0.50	87.20**
Age (years)	-0.04	6.36	0.14	35.56**	-0.04	4.08	0.09	20.05**	-0.04	4.81	0.11	23.07**
Gender (0=male; 1=female)	-0.03	-208.93	-0.15	-37.95**	-0.03	-121.52	-0.09	-15.24**	-0.03	-160.36	-0.11	-21.17**
Migrant status (1=migrant; 0=non-migrant)	0.10	42.92	0.02	5.36**	0.10	102.45	0.05	6.24**	0.10	71.14	0.04	6.43**
Traditional-contemporary index, present residence (+ = more contemporary; - = more traditional	0.30	51.73	0.11	22.57**	0.30	-87.64	-0.19	-13.41**	0.30	42.99	0.09	14.20**
Population pressure/dependent population index, present residence (+ = more pressure; - = less)	-0.21	-13.25	-0.02	-4.22**	-0.21	-6.00	-0.01	-1.72**	-0.21	115.40	0.17	12.12**
Interaction effects, Traditional-contemporary index and												
Educational attainment	na	na	na	na	0.45	9.75	0.15	18.29**	na	na	na	na
Age	na	na	na	na	0.29	3.03	0.25	21.87**	na	na	na	na
Gender	na	na	na	na	0.03	-60.99	-0.08	-13.56**	na	na	na	na
Migrant status	na	na	na	na	0.11	-26.89	-0.03	-3.67**	na	na	na	na
Population pressure index	na	na	na	na	-0.26	-9.43	-0.03	-4.36**	na	na	na	na
Population pressure index and												
Educational attainment	na	na	na	na	na	na	na	na	-0.40	-8.40	-0.08	-10.50**
Age	na	na	na	na	na	na	na	na	-0.20	-2.85	-0.16	-13.77**
Gender	na	na	na	na	na	na	na	na	-0.03	57.24	0.05	8.61**
Migrant status	na	na	na	na	na	na	na	na	-0.10	28.11	0.02	3.31**
Traditional-contemporary index	na	na	na	na	na	na	na	na	-0.26	-10.39	-0.03	-4.38**

Note

(a) Dependent variable is wages in bolivares per month. In all regressions, n = 48,830, the labor force portion of the sample remaining after eliminating cases with missing data. For each regression, in the order presented from left to right, r^2s are 0.32, 0.33, 0.33; Fs are 3,855.12, 3,855.12, 2,150.13.

Significance levels: ** = 0.01, * = 0.05; na = not applicable.

are more important to wages, thus allocating residual variance to place characteristics. Gender differences in labor force participation also are relevant. If employed, women in traditional and/or high population pressure settings receive higher wages than would otherwise be expected, but employment in such locales is considerably less likely. Supportive of this interpretation is Mohan (1986: ch. 10) and International Center for Research on Women (1981). Both find that females permanently in the labor force often exhibit higher than usual motivation and qualifications, leading to better-paid employment.

To further explore gender differences in wages received, models also were estimated for males and females separately. Educational attainment is most important for both, followed by age. With regard to STRUCTURE, however, male wages show a strong direct relationship (i.e., higher in more contemporary settings), whereas female wages are virtually unaffected. An explanation lies in the locational characteristics of each gender's employment opportunities and associated wage levels.

Table 5.3 indicates that approximately 30 percent of the male work force is in Agriculture, with average wages of 351 bolivares per month; other important sectors include Manufacturing (16.3%, 759 bs), Commerce (14.2%, 802 bs), and Public Administration/Defense (10.6%, 858 bs). For females, 67.2% are in Service sectors paying an average of 801 bolivares per month; primarily, Public Administration (9.8%, 858 bs), Health/Community Services (23.4%, 1124 bs), and Domestic/Personal Services (33.3%, 424 bs). Other important sectors of female employment include Manufacturing (14.5%, 759 bs) and Commerce (9.9%, 802 bs). Hence, male employment opportunities with lesser average wages are identified with traditional/rural settings while those with higher average wages are identified with contemporary/urban settings. For females, however, low- and high-wage opportunities are both identified with contemporary/urban settings.

The locational characteristics of each gender are consistent with this argument. Males in the labor force occupy locales spanning a broad segment of the traditional–contemporary continuum, with 40,006 non-migrants having an average STRUCTURE score of +0.70 and 3,657 migrants a +1.93 average, netting +0.80 (Appendix 5.1). By contrast, females occupy a relatively narrow, more contemporary range of the spectrum; 10,981 non-migrants have an average STRUCTURE score of +1.26 and 1,482 migrants a +2.36 average, netting +1.39. Hence, finding that STRUCTURE is less significant for female than for male wages largely reflects the concentration of women, and their employment opportunities, in more contemporary and urban locales.

Concluding observations

In closing this chapter, two sets of observations are offered. One is a synopsis of findings concerning individual labor market experiences in Third World settings and the role therein of personal attributes and place characteristics related to development. The second set of observations concerns general themes highlighted in this book.

Individual labor market experiences: a synopsis of findings

To recapitulate the research design, dependent variables are three labor market experiences – educational attainment, labor force participation, and wages received by individuals as documented by census statistics. Explanatory factors include two multivariate scales measuring place characteristics associated with development at the distrito level, a small political unit. STRUCTURE represents a traditional–contemporary continuum in economic and social structure; PRESSURE represents population pressure within a distrito and the degree to which its population is dependent, rather than economically active. Three equations were estimated. Educational attainment is accounted for by a person's age, gender, and development characteristics of the distrito of *previous* residence. Labor force participation is accounted for by a person's educational attainment, age, gender, and development characteristics of the distrito of *present* residence. Wages attained are accounted for by these same variables and migrant status. In addition, separate analyses of males and females were conducted for labor force participation and wages received.

Educational attainment plays a strong role in wages received by both men and women and in labor force participation by women. Place, or distrito, variation in this direct relationship was minimal for labor force participation, but wage returns to education were greater in contemporary and/or low population pressure settings.

Gender represents the *social category* dimension of this study. Its role in educational attainment was minimal, but women were considerably disadvantaged in labor force participation and somewhat so in wages. Yet, these relationships varied considerably by location. Disparities in educational attainment were less likely, and disparities in labor force participation greater, in traditional, particularly rural locales with higher population pressure. But once in the labor force, women's wages tended to be unaffected by location, while men's were.

Age was broadly significant, but the meaning of this varies. That older people received higher wages reflects skill improvements with age and returns to employment longevity, particularly in contemporary and/or low population pressure settings. But older people also were less educated and less likely to participate in the labor force. This is an artifact of lower development levels in earlier times, which affected educational opportunity, employment availability, and personal attitudes towards taking advantage of these.

The above indicates *place characteristics* associated with development have important *indirect* effects, through personal attributes, on labor market experiences. But *direct* effects also were significant, particularly those of STRUCTURE which represents a traditional–contemporary continuum in socioeconomic characteristics. This variable correlated highly with educational attainment and the nature of employment opportunities in each locale, both of which influence labor force participation and wages received. More generally, labor market experiences are likely to be better in locales that have a greater incidence of contemporary characteristics, a quality closely related to urbanness, and in locales that have less population pressure or are upward transitional. Hence, place characteristics reflecting the spatial spread of development are an integral factor in labor market experiences.

General themes

Chapters Four and Five examine migration and labor market experiences of individuals; and the impact thereon of place characteristics related to development. Both chapters indicate that individuals with identical personal attributes, but residing in different locales, might act dissimilarly because they are faced with different structural conditions, or development contexts. As noted previously, where one lives can have an effect on how one lives.

In the study of labor market experiences, place characteristics associated with development take on even greater relevance. They, as well as social category, affect access to opportunities for educational attainment, labor force participation, and wages received; which in turn impact on personal well-being. Highlighted, therefore, is an *equity* dimension of place characteristics.

These findings for individual behavior also complement Chapter Three's examination of aggregate migration flows. The link between place characteristics related to development and personal behavior underlies aggregate migration, but it is brought to the forefront and elaborated in Chapters Four and Five.

One benefit from examining migration in greater depth, and labor market experiences, is gaining an understanding of mechanisms that contribute to the role of place. Especially prominent in Venezuela were its policies to spatially decentralize educational opportunity and economic activity, which were initiated during the 1950s as part of a broad effort to disseminate petroleum wealth benefits. While educational opportunity was widely improved, raising attainment throughout the country, industrial decentralization was limited primarily to the Ciudad Guayana complex and urban areas in proximity to Caracas, the major city. This difference in the scale of decentralization exacerbated spatial disparities between employment opportunities and the educated work force, and separation from appropriate labor markets raised the individual's risk of being under- or unemployed. The net effect was to increase migration propensities already present in a highly polarized, urban-focussed economy.

These dynamics especially affected females due to a number of factors. First, female educational attainment in less developed, more rural (or peripheral) settings improved more than that of males. Second, economic sectors differ in their tendency to employ a particular gender. Historically, employment opportunities largely associated with males (e.g., in agriculture) were widespread throughout Venezuela, whereas opportunities associated with females (e.g., in service) were more prevalent in urban, more contemporary settings. Efforts to decentralize economic activity had little effect on this long-established pattern. Hence, improved educational attainment brightened employment prospects for both genders, but to fulfill these, urban-directed migration was more integral for females than for males. This gives meaning to the finding in Chapter Four that gender effects on out-migration were considerable in rural and less urbanized distritos, but not elsewhere.

The dynamics outlined above ultimately affect distrito's characteristics; for example, opportunities for education and employment, share of the total population, and population attributes such as average educational attainment, gender ratio, and age distribution. Linking distrito qualities to Venezuela's decentralization policies provides further evidence in support of the perspective that place characteristics associated with development represent the local articulation of world economic and political conditions, donor-nation actions, and national policies. Here, however, the national policy element was examined more by inference than directly, and little attention was given to the means of its articulation in local areas. This is amended in the following two chapters, which focus on specific policies in Ecuador and their role in regional change.

Appendix 5.1 Mean values for variables employed in regression analyses of individual labor market experiences

Variables	All people age 21 or greater	Labor force age 21 or greater				
			Migrants		Non-migrants	
		Total	Female	Male	Female	Male
Individual attributes						
Educational attainment (years of school attended, 0-16)	3.54	3.90	5.81	5.16	5.30	3.32
Labor force participation (%)	52.00	na	na	na	na	na
Wages (in bolivares per month)	na	653.29	682.36	895.56	612.68	640.55
Age (years)	39.37	36.83	30.18	33.24	33.53	38.30
Gender (0=male; 1=female)	0.52	0.22	na	na	na	na
Migrant status (1=migrant; 0=non-migrant)	na	0.09	na	na	na	na
Place attributes[a]						
Traditional–contemporary index for **present** residence (+ = more contemporary distrito; – = more traditional)	1.16	0.93	2.36	1.93	1.26	0.70
Traditional–contemporary index for **previous** residence (+ = more contemporary distrito; – = more traditional)	0.58	0.78	0.49	0.39	1.26	0.70
Population pressure/dependent population index for **present** residence (+ = more pressure in distrito; – = less)	-0.66	-0.59	-1.24	-0.85	-0.79	-0.49
Population pressure/dependent population index for **previous** residence (+ = more pressure in distrito; – = less)	-0.46	-0.54	-0.40	-0.36	-0.79	-0.49
Total persons	167,949	56,126	1,482	3,657	10,981	40,006

Note

(a) Place attributes were obtained by assigning to each individual those principal component scores representing his/her present and previous places of residence. See Chapter Four for a report of the principal component analysis.

na = not applicable.

6 Policy aspects of development and regional change I: Population movements from Ecuador's rural Sierra

Established thus far is the link between exogenous forces outlined in Chapter Two and place characteristics associated with development. Elements of this include the finding that localized development events in Costa Rica and variables defining generalized modernization surfaces in Venezuela are manifestations of such forces. We also saw that Venezuela's decentralization policies were an important mechanism in altering place characteristics associated with development and, therefore, differentiating places.

A second link concerns the prominent role of place characteristics in societal processes such as migration and labor force experiences. This also is demonstrated by the preceding chapters. But elements of the chain − from exogenous forces to place characteristics associated with development to societal process − are based on inference, rather than direct examination.

As a step towards filling this void, Chapter Six considers population movements from Ecuador's rural Sierra during the period 1974−82, giving particular attention to the operation of rural land reform policies. It shows that policy impacts on individuals vary from locale to locale, depending on place characteristics associated with development. We also see that land reform was part of a broader process whereby Ecuador moved from feudal to commodified labor and production structures, a process induced by world economic and political conditions.

Having delineated these links, Chapter Seven focusses directly on socioeconomic change in local areas, using the same time frame but considering all of Ecuador, not just the Sierra. Policies given detailed attention in that chapter pertain to agricultural credit, agricultural pricing, and monetary exchange rates; elements of Ecuador's import substitution industrialization program. Change is delineated; then accounted for by the interaction between policies and place characteristics in 1974. Said another way, place characteristics related to development at a future time reflect the conjunction of earlier characteristics with national policies, world events, and donor-nation actions.

In both chapters, policy is a major concern. Those considered are referred to as *apparently aspatial* in the sense that they lack an areal focus but, nevertheless, have highly variable place-to-place effects. This occurs because the spatial articulation of aspatial policies is *conditioned* by local socioeconomic structures, the elements of which (persons, households, entrepreneurial entities, urban agglomerations, infrastructures, resources, production possibilities, and the like) are located so as to form *areal complexes* that are (more or less) distinct in character. By similar reasoning, *explicitly spatial* policies also exhibit spatial impacts which differ from those anticipated. Examples include Venezuela's policies to spatially decentralize both educational opportunity and economic activity, examined in Chapter Five, as well as regional development programs.

The present chapter's elaboration of these themes is presented in four sections. First, Ecuador and its economic milieu from the 1960s through the early 1980s is described. The second section begins with a brief statement on population mobility and the way it is examined here; attention then turns to a broad-gauged portrayal of Ecuador's agrarian structure, land reform policies, and their effects on movement from the rural Sierra. The third section is concerned with statistical analyses. Indices of agrarian and socioeconomic structure in local political units are reported; then joined with variables representing personal attributes to account for individual out-migration and out-circulation from the rural Sierra. Interpreting these findings employs both the broad-gauged portrayal previously put forth and knowledge of Sierran locales at a more detailed, fine-grained level. The chapter closes by returning to general themes.

Ecuador and its economy

Ecuador, a small South American nation straddling the equator, is comprised of twenty provinces and, in 1974, 114 cantones.[1] These form three regions: from east to west, the Amazon basin or *Oriente*, the Andean Cordillera and its intermontane basins or *Sierra*, and the coastal lowlands or *Costa* (Appendix 6.1).[2]

The Oriente includes eastern slopes of the Andes and Ecuador's portion of the Amazon basin. It comprises more than one-half the national territory, but contains only 3 percent of Ecuador's population. Other features include a fragile tropical rainforest ecosystem, rudimentary levels of infrastructure articulation, and inaccessibility. The Oriente was economically unimportant until 1967 when oil was discovered in the Lago Agrio canton of Napo province, which in turn led to road construction, a pipeline to Quito and Esmeraldas, and agricultural colonization.

The Sierra is a highland area (elevations average approximately 9,500 feet) comprised of two major cordilleras separated by a series of intermontane basins. It contains the largest concentration of Indians, a long-standing settlement system (in part dating from the mid-1500s), and only slightly fewer people than the Costa. Agricultural production, much of which is under traditional systems and for domestic markets, includes barley, beans, potatoes, maize, wheat, beef and dairy cattle. Farming and other economic activities are concentrated in basin areas, which have a greater preponderance of fertile soils, temperate climates, suitable precipitation regimes, and because these characteristics attracted settlement, large urban areas.[3] Ecuador's hacienda system, towards which land reform policies were directed, is primarily associated with this region.

The Costa, which slightly exceeds the Sierra in population, has experienced major economic growth in recent decades, largely due to commercial agricultural production for export markets. Its northern and inland central portions have precipitation levels that support banana, coffee, cotton, oil palm, and sugar cane croppings, as well as dense tropical rainforests. To the south, rainfall declines sharply, leading to semi-arid conditions along the coast from Portoviejo-Bahia to the Peruvian border. However, commercial fishing flourishes, and inland is Ecuador's most productive agricultural area, a series of alluvial plains where rainfall is abundant and soils fertile. These include the Guayas basin (in proximity to Ecuador's largest city, Guayaquil) which leads in banana, cacao, and rice production.

Overall, Ecuador's space-economy changed significantly since the mid-1950s. Factors underlying this include 1964 and 1973 Agrarian Reform legislation, discussed in the following section; continuing strength of export agriculture in the Costa; and population shifts related to both occurrences. Changes also were triggered by petroleum discoveries in 1967, which initiated a period of unprecedented economic growth. Reliance on commercial agriculture was supplanted as petroleum products increased from 18.6 percent of total exports in 1972 to 46.5 percent in 1977 (World Bank 1979: 255); government revenues grew dramatically; and import substitution industrialization policies channeled resources to industrial development, often at the expense of agriculture.

After 1979, economic growth slackened due in large part to worldwide recessionary conditions and declining petroleum prices/demand (World Bank 1984: 5–7). Ecuador's petroleum production/export levels dropped, foreign exchange constraints grew, and terms of trade became increasingly unfavorable. These conditions, as well as uncertainties related to border conflicts with Peru and a change in government (Hurtado 1985; Schodt 1987), led the private sector to slow its rate of investment. In 1982–83, already stagnating agriculture was further weakened by droughts and floods that destroyed crops and marketing infrastructures in the Costa, thus increasing reliance on food imports and need for scarce foreign exchange. There were, however, exceptions to this trend. The volume of shrimp exports, for example, increased by 48 percent annually from 1979 through 1983, primarily due to pisciculture introduced to El Oro and Guayas provinces during the 1970s (World Bank 1984: 15–17).

Agrarian structure, land reform, and movement from the rural Sierra: a broad-gauged portrait[4]

With this brief description of Ecuador as background, attention turns to out-migration and out-circulation from rural portions of the Sierra, the region most affected by land reform policies. Of particular concern is the role of land reform in population mobility, a major problem of the Third World (Standing 1985a). Internal movements have depopulated some areas, overcrowded others, and in both instances, disrupted prevailing socioeconomic systems. These shifts may be set off by cataclysmic events such as earthquakes, desertification, internal warfare, or a sharp rise/fall in world commodity prices. Often, however, massive population shifts are the (unanticipated) outcome of government actions, as discussed generally by Findley (1981), Rhoda (1983), Shrestha (1987), and in Chapter Five for Venezuela.

Two forms of mobility are commonly identified; permanent migration and circulation. The latter, which has received considerably less study than migration, is defined by Zelinsky (1971: 255–6) as

a great variety of movements, usually short-term, repetitive, or cyclic in nature, but all having in common the lack of any declared intention of a permanent or long-lasting change in residence.

Circulation thus includes both seasonal and sporadically timed moves for purposes of temporary employment and, according to some typologies, daily commuting; more generally, these movements involve geographic separation of kinship obligations and work-related activities (Conaway 1976, 1977; Standing 1985b). Circulation may be a long-established (cultural) tradition, or a contemporary response to changing conditions. Among the latter are short-term movements by individuals as part of a family (or community) strategy for coping with decreases in its (per capita) resource base, a practice particularly prevalent in rural areas where population increases have been considerable, land availability is fixed or diminishing, and alternative employment opportunities are located elsewhere. Ecuador exemplifies such a setting.[5]

This study of out-circulation and out-migration centers on the operation of Ecuador's land reform policies among persons residing in rural portions of Ecuador's Sierran region during the period 1974–82. A critical ingredient is detailed knowledge of those policies and of Ecuador's socioeconomic fabric, particularly at the local level; i.e., a *sense of* Ecuador and its subdivisions as *places*. Combining this with statistical analyses demonstrates that policy effects (or impacts) on socioeconomic conditions and individual behavior vary from locale to locale, depending on place characteristics related to development.

To specify the role of place, personal attributes also are considered as are differences in out-circulation and out-migration processes. Concerning the latter, examining circulation and migration within the same population is rare in itself, and when done, differences in process have been given scant attention.[6]

As a first step in the investigation, a broad-gauged portrait of agrarian structure and land reform is presented. Later, to interpret statistical findings, this is augmented by more detailed, finer-grained knowledge of Sierran locales.

Agrarian structure

Agriculture employs a significant portion of Ecuador's labor force (nearly one-half in 1982), constitutes a large segment of its exports (nearly one-quarter in 1983, petroleum being the other major export), provides inexpensive food for urban/industrial enterprise, and is spatially more dispersed than other economic activities (Commander and Peek 1986; World Bank 1984: 8–10, 77, 126–7). Further, different forms of production and labor systems are *distinctly linked to* particular *regions.*

Rural landholding is characterized by *latifundios*, or large holdings, that occupy most of the land, and *minifundios*, or small holdings, that contain most of the rural population (Odell and Preston 1978; ch. 3). Among latifundios, the traditional form of tenancy is family estates, or *haciendas.* Found primarily in the Sierra, these produce largely for domestic markets, often with labor-intensive methods. Until land reform in the mid-1960s, haciendas were characterized by feudalistic tenancy systems such as the *huasipungo.* Peasants worked several days a week throughout the year, including slack seasons when road construction, fence repairs, and other non-agricultural tasks were undertaken; female *huasipungueras* often served in the *hacendado*'s house. In return, peasants were given the right to farm small land plots, gather wood, and graze cattle within the hacienda; they also were assisted by the hacendado in instances of extreme need such as illness, a bad harvest, or legal entanglements. Hence, this *patron* system freed laborers from severe deprivation, but kept them dependent on the landowner (Handelman 1980; Phillips 1987).

Alternatives to huasipungo tenancy included *arrendamiento*, renting hacienda land for cash, and *aparceria*, for payments in kind such as crops or a specified number of work days per week. These became more common after land reform, when the huasipungo was officially abolished (Commander and Peek 1986; Griffin 1976); a response that maintained high levels of social dependency. Also common, before and after land reform, were landless peasants (*arrimados*) who labored for wages but were tied both economically and socially to the hacienda system.[7]

Minifundios constitute the other major landholding form of the Sierra. Semi-subsistence agriculture is common; i.e., crops are largely for home consumption with surplus sold in domestic markets. Size ranges from one-eighth to five hectares, but is trending lower due to partible inheritance practices whereby land is subdivided among all heirs. This and population growth have increased land-use intensity, which together with traditional methods of cultivation accelerates soil erosion and reduces productivity. The situation is further aggravated because Sierran minifundistas often hold land without title or rent under an arrendamiento or aparceria arrangement; being excluded, therefore, from policy benefits requiring proof of ownership, such as agricultural credit (Lawson 1986, 1988). More generally, national development policies are biased towards contemporary production systems, thus depriving Sierran minifundios of benefits available to other types of landholding. These various factors create a need for supplementary income, which generally is obtained through circulation or migration (Commander and Peek 1986).

Also relevant to the present study are industrial plantations. Located predominantly in the Costa, these latifundios tend to be smaller than haciendas and characterized by wage labor, production practices that are more capital/technology-intensive, orientation towards export markets, and, frequently, foreign ownership. Crops include bananas, coffee, cacao, sugar, cotton, and the like. Labor needs are in large part seasonal, and there is no hacienda-type social relationship between workers and owners. Hence, temporary employment is common, and an important element of Ecuador's circulation system.

Land reform and its impacts

Agrarian reform laws activated in 1964 (and subsequent years) addressed adverse aspects of Ecuador's landholding system, particularly those associated with the hacienda system. But the result was not entirely favorable for peasants. In 1954, farms of less than five hectares constituted 81.7 percent of all holdings and 11.4 percent of the farmland; in 1974, these percentages were 77.1 and 12.8, respectively. There was little change because large landholders and industrialists were able to shift the law's emphasis to favor output and production efficiency over equity. As a result, expropriation could be avoided by improving production techniques, cultivating more land, and abolishing feudalistic tenancy systems; but some simply divided land among family members to achieve holding levels below that requiring expropriation (Blankstein and Zuvekas 1973; Commander and Peek 1986: 81–4; Griffin 1976; Handelman 1980; Peek 1982).[8]

The outcome was minimal change in land distribution, but considerable modification to the structure of production. Larger portions of haciendas were cultivated, and capital-intensive techniques introduced. Contributing to this transformation was a significant increase in agricultural credit, much of which was invested to expand livestock and dairy farming on large landholdings, and to modernize crop cultivation on medium-size units (Commander and Peek 1986).

These changes were accompanied by expulsion of peasants from haciendas, where huasipungueros decreased from an estimated 100,000 in the 1950s to about 2,600 in 1974 (Commander and Peek 1986: 81). Although many (perhaps the majority) of these peasants became landowners, they received less desirable sites that averaged only 3.5 hectares in size, below the 5-hectare minimum established by law (Blankstein and Zuvekas 1973: 82).

More generally, agrarian reform worsened the position of peasant agriculturalists in several ways. First, abolition of huasipungo voided their right to gather wood and graze cattle on haciendas, and landowners were no longer compelled to assist in times of need. Second, land given to peasants was often less fertile than that previously cultivated. Third, minifundios became even smaller, decreasing on average from 1.61 hectares in 1954 to 1.45 in 1974 (Peek 1980: 617). Fourth, small farm productivity stagnated or declined due to increasing population pressure, deteriorating soil quality, and macroeconomic/policy conditions that adversely affected domestic agriculture (Lawson 1986, 1988).

Due to these circumstances, peasants became increasingly dependent on employment external to their landholding, but simultaneously, production mode changes related to land reform reduced demand for agricultural labor. Mechanization was partly responsible; also, many large landholdings (particularly ones near urban centers) took up dairying, which tends to be less labor-intensive than alternative endeavors. Economic rationality, which became an important component of landholder decision-making, also contributed in that financial obligation could be reduced by replacing permanent workers with temporary ones, an alternative facilitated by labor commodification (Commander and Peek 1986: 84–7; Peek 1982). In 1974, therefore, hired-in labor was largely temporary for all farm sizes and constituted more than 50 percent of the total labor input on farms over 50 hectares; a change in labor use that considerably increased the incidence of circulation in agriculture (Commander and Peek 1986: 85).

With regard to other economic sectors, labor demand in activities related to petroleum exploitation grew considerably in the 1970s, but primarily on a short-term basis for construction of pipelines, roads, buildings, and the like. Urban employment expanded in absolute terms, but more slowly than the urban labor force, giving rise to severe underemployment. In addition, the number of urban workers earning below minimum wage increased, and the poor's share of total income declined (Commander and Peek 1986; Peek 1979: 5–12). Finally, these conditions worsened considerably after 1979 when Ecuador's economy turned downward. In general, then, uncertainty characterized employment prospects in non-agricultural sectors, increasing risks associated with permanent migration and leading many to circulation strategies, particularly peasants with arable land.[9]

Statistical analyses

Having examined broad effects of Ecuador's agrarian structure and land reform policies on movement from the rural Sierra, attention now turns to statistical analyses for a more rigorous and detailed perspective. This is presented in three parts. First, indices reflecting the Sierra's agrarian situation at the canton level are delineated. Reported next is the strategy by which place indices and variables representing personal attributes are joined to account for individual out-migration and out-circulation. Third, findings are explicated, giving particular attention to the role of place characteristics in spatially differentiating policy effects on out-circulation and out-migration.

Indices of structural conditions in Ecuador cantones

Conditions discussed in the preceding section are encapsulated in canton indices, the basis of which is two principal components analyses for the *whole* of Ecuador, reported in Brea (1986). One analysis pertains to agrarian structure, the other to the socioeconomic environment. After describing each, their application to the rural Sierra is delineated.

INDICES OF AGRARIAN STRUCTURE

Twelve variables pertaining to cantones, taken from Ecuador's 1974 Census of Agriculture, were combined into three indices

(Table 6.1). The first, termed TIME, distinguishes cantones with long-standing holdings (+) from ones with more recent (−) establishments. Long-standing agricultural settings are characterized by a high percent of farm area owned through title, high population pressure, and a high percent of farm area in the small-size category. More recently established settings are characterized by a high percent of farm area held through agrarian reform or colonization, high percent of farm area held without title, and farms in the medium-size category that tend to be large compared to other farms in that category. Locationally, long-standing agricultural settings are found among cantones in the Sierra, while recently developed settings are found in the Oriente and northern Costa where spontaneous settlement and colonization has been significant since the 1950s.

The second index of agrarian structure, termed PRODUCTION ORIENTATION, distinguishes cantones where large segments of the population are engaged in semi-subsistence (+) agriculture from ones where production tends to be for domestic and/or export markets (−). Semi-subsistence-oriented settings are characterized by a high proportion of agricultural land under annual crops; also by high population pressure and a high percent of farm area in the small-size category. Market-oriented settings are characterized by a high percent of farm area in the medium-size category, farms in the small- and medium-size categories that tend to be large compared to other farms in those categories, and a high percent of farm area owned through agrarian reform or colonization; these settings also would have a high proportion of land under perennial crops. Locationally, cantones distinguished by semi-subsistence production are in the Sierra; while cantones distinguished by market production are in the Costa, as would be anticipated by the preceding description of Ecuador's space-economy.

The third index of agrarian structure, termed SIZE, distinguishes cantones where much of the farm area is taken up by large farms (+) from those where medium- and small-size holdings (−) prevail. Settings with large agricultural enterprises are characterized by a high percent of farm area in the large-size category, and by farms in the large-size category that tend to be large compared to other farms in that category. Settings with a mixture of medium and small holdings are characterized by a high percent of farm area in those size categories and a high percent of farm area rented through either sharecropping or cash. Locationally, large-

Table 6.1 Principal components analysis of agrarian variables for Ecuador cantones in 1974, varimax rotation

Variable	Mean	I	II	III	Commu nality
			Component		
Percent of farm area in small holdings (less than 5 ha.)	11.96	0.48	0.70	-0.20	0.76
Percent of farm area in medium-size holdings (5-50 ha.)	29.60	0.23	-0.60	-0.56	0.72
Percent farm area in very large holdings (more than 500 ha.)	24.58	0.23	0.08	0.90	0.87
Mean size of small farms (ha.)	1.81	-0.11	-0.85	-0.06	0.73
Mean size of medium farms (ha.)	15.60	-0.82	-0.27	-0.15	0.77
Mean size of very large farms (ha.)	1527.66	0.11	0.15	0.73	0.56
Proportion of area under annual crops (annual crops/(annual + perennial crops))	0.63	0.13	0.75	0.21	0.62
Percent of farm area owned with title	71.08	0.91	-0.02	-0.03	0.84
Percent of farm area owned through agrarian reform or colonization	6.73	-0.52	-0.30	0.10	0.37
Percent of farm area rented (cash or sharecropping)	3.02	0.16	0.13	-0.48	0.27
Percent of farm area held without title	11.11	-0.86	-0.04	-0.11	0.75
Rural population pressure (rural population/cultivated land)	43.69	0.44	0.48	-0.22	0.47
Percent variance explained – by each factor		33.1	18.4	12.9	
– cumulative		33.1	51.5	64.4	

n = 111

scale agricultural settings are found in both Costa and Sierra cantones. The preceding description of Ecuador's space-economy suggests the Costa portion tends to be export-oriented commercial agriculture; the Sierra portion, cattle and dairy farming. Settings characterized by medium- and small-size holdings have no regional identification, being found throughout Ecuador.

INDICES OF THE BROADER SOCIOECONOMIC ENVIRONMENT

Concerning the socioeconomic situation within which agrarian structure is imbedded, twenty-one variables pertaining to cantones were combined into three indices (Table 6.2); data are from Ecuador's 1974 Census of Population. Only the first principal

Table 6.2 Principal components analysis of socioeconomic variables for Ecuador cantones in 1974, varimax rotation

Variable	Mean	Component I	Component II	Component III	Commu nality
Mean age	22.06	-0.10	0.84	0.14	0.74
Dependency ratio ((population<15 + population>65)/population aged 15-65)	1.00	-0.32	-0.25	-0.64	0.57
Percent male	51.14	-0.02	-0.80	0.43	0.83
Percent employed in agricultural sector	67.76	-0.76	-0.52	-0.26	0.91
Percent employed in mining sector	0.44	0.10	-0.17	0.74	0.58
Percent employed in manufacturing sector	9.32	0.13	0.82	0.12	0.71
Percent employed in utilities sector	0.26	0.48	0.21	0.04	0.28
Percent employed in construction sector	2.97	0.51	0.19	0.59	0.64
Percent employed in tertiary activities	7.75	0.87	0.14	-0.18	0.82
Percent of population residing in rural areas	78.35	-0.87	-0.18	0.12	0.81
Mean years of education	2.83	0.88	0.00	0.27	0.84
Percent of population with no education	32.31	-0.66	0.15	-0.19	0.50
Percent of population with college education	0.74	0.84	0.23	0.20	0.79
Population pressure (total population/ employed population)	3.61	0.16	-0.15	-0.67	0.49
Child-woman ratio (population under 5/ females aged 15-49) x 1000	826.80	-0.36	-0.74	-0.03	0.68
Percent of employed who are agricul- turalists or fishermen	68.44	-0.76	-0.56	-0.20	0.93
Percent of employed who are professionals, managers, or technicians	4.04	0.88	0.03	0.19	0.81
Percent of employed who are clericals	1.89	0.90	0.17	0.17	0.86
Percent of employed who are vendors or sales persons	5.02	0.81	0.07	-0.27	0.73
Percent of employed who are craftsmen or operators	12.58	0.29	0.82	0.19	0.80
Percent of employed persons who are in service occupations	3.78	0.92	0.18	0.18	0.92
Percent variance explained – by each factor		47.1	15.1	10.3	
– cumulative		47.1	62.2	72.5	

n = 111

component, termed MODERNITY, is deemed relevant to circulation and migration movements from rural areas of the Sierra; this represents a traditional–contemporary continuum in the socioeconomic structure of Ecuador's cantones. More traditional settings, those with negative (–) principal component scores, are characterized by a high percent of the population residing in rural areas, high level of employment in agriculture, high ratio of children to women, and a low level of educational attainment. More contemporary settings, those with positive (+) principal component scores, are characterized by high levels of employment in utilities, construction, tertiary, and service activities; and a high level of educational attainment. Locationally, more contemporary cantones contain, or are proximate to, sizable urban agglomerations such as Quito, Cuenca, Ambato, Riobamba, and Loja in the Sierra and Guayaquil in the Costa (Appendix 6.1). More traditional cantones are found in spatially remote and/or rural portions of the Sierra and in the Costa's Guayas Basin and province of Manabi.

ARTICULATION OF STRUCTURAL INDICES IN THE SIERRAN SETTING

Because the preceding indices were derived for all of Ecuador, their meaning in the Sierran context is an important issue. Elaborating this draws on the chapter's broad-gauged description of land reform and socioeconomic conditions in the rural Sierra; Programa Nacional de Regionalizacion Agraria's (1979) highly detailed accounts of Ecuadorian locales, often at the sub-canton level; and component scores for Sierran cantones (Table 6.3). Canton locations are shown in Appendix 6.1.

Cantones characterized by the hacienda tradition have strongly positive scores on SIZE, having retained their large farm trait even though land reform created numerous minifundios. Sizable landholdings often are remnants of haciendas; as in Cayambe (+1.52) in Pichincha, Alausi (+1.33) and Guamote (+3.45) in Chimborazo. But large farms also were assembled by *comerciantes* through buying land from hacendados or minifundistas; exemplifying this are Mejia (+1.59) in Pichincha, Salcedo (+1.35) in Cotopaxi, Patate (+1.42) in Tungurahua, Biblian (+1.00) in Canar.

Concerning TIME, the above cantones should have positive scores because the hacienda tradition is long-standing, large farms or remnants of haciendas occupy a sizable portion of total land area, and their title to ownership is well established. While this expectation is realized, TIME scores for hacienda-tradition cantones tend to be modest in magnitude. Higher positive scores are found among cantones characterized by economic marginality, under-utilized hacienda or latifundio land, and long-standing medium- and small-size farms. Taken together, this suggests low levels of commercial interest regarding rural land. Examples

Table 6.3 Indices pertaining to agrarian and socioeconomic structure, out-circulation, and out-migration from the rural Sierra of Ecuador

Cantones	Provinces	Agrarian structure			Socioeconomic environment	Indices[a]	
		TIME	PRODCT'N ORIENT'N	SIZE	MODERNITY	circ' tn	migr' tn
Tulcan	Carchi	0.00	0.27	0.50	1.42	15.5	39.3
Espejo		0.12	0.09	0.04	-0.34	20.3	49.1
Montufar		0.65	0.37	-0.80	0.33	12.4	51.6
Ibarra	Imbabura	0.38	0.55	0.72	0.87	26.2	54.0
Antonio Ante		md	md	md	0.26	41.7	39.9
Cotacachi		-0.03	0.72	0.49	-0.90	50.0	42.4
Otavalo		0.43	1.19	-0.06	-0.35	40.3	38.5
Cayambe	Pichincha	0.30	0.65	1.52	0.02	32.8	43.4
Mejia		-0.14	0.83	1.59	0.50	23.8	40.8
Pedro Moncayo		0.46	0.50	0.44	-0.58	33.5	56.9
Ruminahui		0.22	1.17	0.27	1.67	19.7	33.6
Santo Domingo		-0.65	-1.29	-0.42	0.36	33.1	100.0
Latacunga	Cotopaxi	0.05	0.80	0.41	0.06	17.9	41.4
Pangua		1.03	-1.46	-0.14	-0.58	6.9	65.0
Pujili		0.31	-0.15	0.55	-1.11	29.0	37.1
Salcedo		0.30	0.95	1.35	-0.62	29.3	36.6
Saquisili		md	md	md	-0.03	19.3	36.2
Banos	Tungurahua	0.07	0.40	0.36	0.82	46.2	89.4
Pelileo		md	md	md	-0.69	20.4	65.2
Pillaro		md	md	md	-0.30	31.4	42.2
Patate		0.72	0.21	1.42	-0.72	27.2	51.2
Quero		md	md	md	-1.12	21.7	46.3
Guaranda	Bolivar	0.55	-0.32	-0.61	-0.31	11.7	52.5
Chillanes		0.78	-0.78	-1.82	-0.86	19.7	67.9
Chimbo		1.05	-0.69	-1.71	-0.25	16.2	60.4
San Miguel		0.98	-0.21	-1.76	-0.29	12.8	66.1
Riobamba	Chimborazo	0.53	0.96	0.51	0.92	20.0	52.5
Alausi		0.31	0.13	1.33	-0.70	86.9	62.1
Colta		0.58	0.12	-0.61	-1.46	29.7	46.1
Chunchi		0.55	0.72	-0.78	-0.53	26.6	58.4
Guamote		0.71	-0.21	3.45	-1.68	34.1	42.3
Guano		0.74	0.91	-0.44	-1.01	24.8	41.5
Azogues	Canar	0.57	1.23	0.63	-0.30	15.2	39.4
Biblian		0.27	1.21	1.00	-0.85	12.8	41.8
Canar		-0.16	0.21	0.61	-0.57	21.0	29.2
Giron	Azuay	0.83	0.62	-0.58	-0.99	79.0	63.9
Gualaceo		md	md	md	-1.13	37.6	68.0
Paute		md	md	md	-0.86	31.7	73.3
Santa Isabel		md	md	md	-0.73	20.7	55.0
Sigsig		0.62	0.90	0.31	-1.33	100.0	77.2
Loja	Loja	0.58	0.16	0.10	1.50	49.0	75.9
Calvas		0.07	0.03	-0.24	-0.16	27.9	75.6
Celica		0.80	-0.23	-0.64	-0.15	17.9	75.4
Espindola		-0.62	0.05	0.78	-0.94	25.2	74.0
Gonzanama		0.23	0.05	0.42	-0.59	32.8	74.3
Macara		0.24	0.11	0.29	0.36	47.6	91.0
Paltas		0.76	-0.64	-0.46	-0.65	27.9	74.3
Puyango		0.88	-0.66	-1.06	-0.52	19.0	77.3
Saraguro		1.09	0.22	1.70	-1.11	47.6	73.2

Note

(a) Standardized indices for out-circulation were derived by assigning a 1.0 value to the highest rate; computing the ratio of other out-circulation rates to the highest rate; then using that ratio as each canton's standardized index. A similar procedure was followed for out-migration.

md = missing data.

include Chillanes (+0.78, with a SIZE score of -1.82), Chimbo (+1.05, -1.71), and San Miguel (+0.98, -1.76) in Bolivar; Giron (+0.83, -0.58) in Azuay; Celica (+0.80, -0.64), Paltas (+0.76, -0.46), and Puyango (+0.88, -1.06) in Loja.

Negative scores on TIME indicate more recent settlement, through spontaneous means or colonization that began in the 1950s. Among Sierra cantones, only Santo Domingo (-0.65) in Pichincha and Espindola (-0.62) in Loja stand out. Espindola represents a situation of considerable class conflict, aided by the canton's remote location. Recent settlement occurred on uncultivated and/or uninhabited land, much of which had been claimed by haciendas/latifundios but was not developed or utilized. This description also applies to portions of other cantones; among those mentioned earlier, Chillanes, Chimbo, and San Miguel in Bolivar; Pangua in Cotopaxi; Celica, Paltas, and Puyango in Loja.

Positive scores on PRODUCTION ORIENTATION indicate cantones with a high proportion of land area devoted to annual, rather than perennial, crops (e.g., onions, barley, potatoes, beans, maize). This is associated with minifundios in intermontane segments of the Sierra that often evolved out of, or side-by-side with, the hacienda tradition. Semi-subsistence agriculture is usual, but it typically includes some production for domestic markets. Artisannal activity and/or temporary employment also are important sources of income. Exemplary cantones include Otavalo (+1.19) in Imbabura; Cayambe (+0.65), Mejia (+0.83), and Ruminahui (+1.17) in Pichincha; Salcedo (+0.95) in Cotopaxi; Biblian (+1.21) in Canar.

Negative scores on PRODUCTION ORIENTATION indicate a high proportion of perennial crops such as bananas, citrus, cocoa, coffee, fruits, and sugar cane. In the Sierra context, these may be for either market or semi-subsistence purposes. Perennials are grown in suitable microclimatic regimes of intermontane areas and in cantones at the Sierra's eastern or western fringe. *Fringe cantones* often include lower altitude areas on the Costa or Oriente side of the Andean Cordillera where latifundio/hacienda control was weak or land uninhabited. Recent settlement is common; many landholdings are in the small- and medium-size categories but tend to be large for those categories; and commercial agriculture with hired in labor is widespread. Examples of fringe cantones with negative scores on PRODUCTION ORIENTATION include Santo Domingo (-1.29) in Pichincha, Pangua (-1.46) in Cotopaxi, Chillanes (-0.78) and Chimbo (-0.69) in

Bolivar, Paltas (-0.64) and Puyango (-0.66) in Loja.

In terms of MODERNITY, positive scores are found for Sierran cantones containing urban agglomerations; for example, Tulcan (+1.42) in Carchi, Ibarra (+0.87) in Imbabura, Ruminahui (+1.67) (adjacent to Quito) in Pichincha, Banos (+0.82) in Tungurahua, Riobamba (+0.92) in Chimborazo, Loja (+1.50) in Loja.[10] Negative scores, indicating more traditional socioeconomic settings, are found in all (but one) non-urban cantones south of Ambato. Locational remoteness is partly responsible; cantones in Loja, the southernmost province, provide an example. But a more general observation is that contemporary development in the rural Sierra has been centered on Quito. Within this sphere, from Ambato to the Colombian border, traditional socioeconomic settings are indicated only for distinct situations. One is 'fringe' cantones that have a portion of their area on the Costa or Oriente side of the Cordillera; specifically, Cotacachi (-0.90) in Imbabura and Pangua (-0.58) in Cotopaxi. A second situation is cantones where dissolution of the hacienda tradition began decades before land reform, often accompanied by class conflict, leading to minifundio-based communities, sometimes long-standing; specifically, Pedro Moncayo (-0.58) in Pichincha, Pujili (-1.11) and Salcedo (-0.62) in Cotopaxi. While these examples pertain to areas in the vicinity or north of Ambato, similar situations are found among cantones to its south.

Linking place indices and variables representing personal attributes

Canton indices are used in conjunction with data on individuals provided by a CELADE sample of 834,790 records from Ecuador's 1982 Census of Population.[11] Application of the following criteria reduced these to 32,888 records for analyzing circulation and 40,395 for migration.[12] First, only the economically active are included, i.e., persons indicating employment who were twelve or more years old, the age criterion used in asking occupation/employment questions. Second, non-migrants and non-circulators are included only if their canton of present residence is in the *rural Sierra*; migrants and circulators only if their origin canton is in the *rural Sierra*.[13] Third, migrants include only those moving within nine years preceding the census.[14]

Circulator/non-circulator status (OUTCIRC$_i$) and migrant/non-migrant status (OUTMIG$_i$) are related to an individual's gender (GENDER$_i$), age (AGE$_i$), marital status (MARITAL$_i$), educational attainment (EDUC$_i$), and each index pertaining to his/her canton of *previous* residence (TIME$_i$, PRODUCTION ORIENTATION$_i$, SIZE$_i$, MODERNITY$_i$).[15] The analytical procedure is logistic regression. As discussed in Chapter Four, this may be likened to linear regression with a dichotomous dependent variable indicating whether each sampled individual i is an out-migrant/circulator (1) or stayer (0).[16] In fact, however, the dependent variable is a ratio of the probability of moving (Pmov$_i$) to the probability of staying (1−Pmov$_i$), logarithmically transformed. Taking OUTCIRC$_i$ to represent this ratio for out-circulation, direct relationships were estimated for the variable set overall, i.e.,

$$OUTCIRC_i = f(GENDER_i, AGE_i, MARITAL_i, EDUC_i,$$
$$TIME_i, PRODUCTION\ ORIENTATION_i,$$
$$SIZE_i, MODERNITY_i)$$

and similarly for OUTMIG$_i$. Also estimated were zero-order relationships; that is, for each independent variable by itself against OUTCIRC$_i$ and OUTMIG$_i$. Using AGE as an example,

$$OUTCIRC_i = f(AGE_i)$$

Findings

Table 6.4 shows mean values on each independent variable for non-circulators, circulators, non-migrants, and migrants; zero-order logistic regression relationships for OUTCIRC$_i$ and OUTMIG$_i$; and multiple variable logistic regression relationships. Taken together, these statistics indicate the following.

Concerning personal attributes, circulators and migrants tend to be males and persons with a less attached marital status, but in differentiating each with non-movers, the tendency is more marked for circulators. Circulators and migrants also tend to be better educated, but here, in differentiating each with non-movers, the tendency is more marked for migrants (and considerably so). Also, migrants tend to be noticeably better educated than circulators. Finally, circulators and migrants are likely to be younger than stayers, with differentials of similar magnitude.

Table 6.4 Mean values and logistic regressions related to out-circulation and out-migration from the rural Sierra of Ecuador

Independent variables	Mean values[a]				Out-circulation[a] — Logistic regressions[a]					Out-migration[a] — Logistic regressions[a]				
	Non-circ's	Circ's	Non-mig's	Mig's	Zero order		Multiple variable			Zero order		Multiple variable		
					b	[a]	b	Beta[a]	[a]	b	[a]	b	Beta[a]	[a]
Gender (1=male; 2=female)	1.22	1.13	1.20	1.19	-0.30	-10.98	-0.34	-0.51	-11.35	-0.03	-2.95	-0.04	-0.05	-3.84
Age (years)	36.86	31.50	36.27	31.26	-0.01	-17.48	-0.01	-0.71	-13.89	-0.01	-37.51	-0.01	-0.30	-25.40
Marital status (1=sngle; 2=separated; 3=mrrd, union)	2.29	2.09	2.28	2.14	-0.11	-11.48	-0.06	-0.20	-4.68	-0.08	-17.07	0.01	0.02	1.30*
Educational attainment (years attended, 0-19)	5.00	5.20	4.53	5.65	0.01	2.44	0.00	0.03	0.60*	0.04	31.17	0.03	0.22	20.06
TIME: long-standing(+) vs recent(−) ag settlement	0.32	0.34	0.38	0.36	0.07	2.83	0.08	0.13	3.03	-0.07	-6.35	0.01	0.01	0.46*
PRODUCTION ORIENT'N: semi-sbst(+) vs mkt(−)	0.27	0.30*	0.38	0.14	0.02	1.58*	0.01	0.02	0.41*	-0.30	-40.23	-0.30	-0.39	-31.45
SIZE: large(+) vs medium/small(−) size farms	0.22	0.29	0.27	0.09	0.05	4.29	0.06	0.16	4.00	-0.13	-22.62	-0.04	-0.06	-4.99
MODERNITY: contemp(+) vs trad(−) socioec struct	-0.05	-0.12	-0.15	-0.10	-0.05	-4.16	-0.05	-0.14	-3.55	0.04	6.89	0.05	0.08	7.43

Out-circulation: n circulators 2,664 (8.1%); n stayers 30,224 (91.9%); n total 32,888. Log likelihood ratio = 604.5[b]

Out-migration: n migrants 19,794 (49.0%); n stayers 20,601 (51.0%); n total 40,395. Log likelihood ratio = 3413.5[b]

Notes

(a) For all Non-circulator–circulator and Non-migrant–migrant comparisons, mean values are significantly different from one another at the 0.01 level or better, except the comparison noted by * is not significant at the 0.05 level.

(b) Logistic regression analyses were done with the SPSS program PROBIT. Separately, t-values were estimated as the b-coefficient divided by its standard error; Betas were estimated as b multiplied by the ratio of the standard deviations of the independent and dependent variables (S_i/S_y); log likelihood ratios were computed by the method outlined in Aldrich and Nelson (1984: 55–6). All t-values are significant at the 0.01 level, except where indicated by *; all log likelihood ratios are significant at the 0.01 level.

Whereas each personal attribute has a similar effect on circulation and migration (e.g., age varies inversely with both), this is *not* so for place characteristics. Circulation is more likely from cantones with long-standing agricultural holdings, semi-subsistence agriculture, large farms, and characteristics indicating traditional socioeconomic settings. Migration is more likely from cantones with relatively recent agricultural holdings, a market orientation in agricultural production, medium- and small-size farms, and characteristics indicating contemporary socioeconomic settings.

The remainder of this section addresses the role of place characteristics in greater detail, focussing first on circulation, then on migration. It closes with general observations on the role of land reform policies in movement from the rural Sierra. These discussions draw together the chapter's broad-gauged portrait of Ecuador's agrarian structure and land reform policies, its more detailed canton-level accounts of structural conditions in the rural Sierra, and statistical findings. Among the latter are standardized rates of out-circulation and out-migration (Table 6.3), converted to quintiles and mapped (Figure 6.1) to indicate spatial patterns.[17]

CIRCULATION

Circulation is associated with minifundios. Partible inheritance, increasing population pressure related to high rates of population growth, soil erosion, limited borrowing power, and the like have reduced the minifundio's ability to support family units. As a result, minifundistas also undertake artisannal endeavors and/or temporary employment, activities that generally transpire in other locales. Choosing circulation, rather than migration, reflects ties to both land and culture.

Concerning land, even meager agricultural production contributes to family unit needs and, if there is a marketable surplus, may provide supplemental income. Also, Programa Nacional de Regionalizacion Agraria (1979) refers to many communities where members left to establish a farm or occupation elsewhere, but returned after some years. Knowledge of this, together with fewer wage-labor opportunities after Ecuador's economic downturn in the late 1970s, is further inducement to maintain the safety net of one's land.

Concerning cultural ties, these are most commonly associated with minifundios growing out of the hacienda/huasipungo system.

But Programa Nacional de Regionalizacion Agraria indicates they also are associated with long-standing minifundio locales where closely-knit, independent Indian communities have resisted incursions by latifundistas and others, both in earlier decades and currently. In such communities, furthermore, circulation is often an established tradition, whereas it may be more recent among huasipungueros.

To embellish this picture, consider Figure 6.1 (together with Appendix 6.1 and Table 6.3). One pocket of out-circulation is the cantones of Sigsig and Giron in Azuay and Saraguro in Loja. These are characterized by independent Indian communities, highly deteriorated land conditions, minifundio (often microfundio) agriculture, a history of conflicts both between and within socio-economic classes, and circulation as a long-standing tradition (Programa Nacional de Regionalizacion Agraria 1979: Documento B, 127–40; Belote and Belote 1985). Another pocket of out-circulation is Antonio Ante, Cotacachi, and Otavalo in Imbabura, where minifundio conditions have become highly deteriorated; among Otavalenos, there also is a tradition of temporary move-ment related to their widely known artisannal and merchandising abilities (Programa Nacional de Regionalizacion Agraria 1979: Documento B, 52–64). Alausi in Chimborazo represents a high out-circulation canton where the hacienda/huasipungo system was well established. While a similar observation applies to Loja in Loja, more interesting is the extreme drought it experienced in 1968. This led to considerable out-migration, transformation of social class relationships, class conflicts, and changes in the structure of production; Loja's high rate of out-circulation is derivative of these occurrences (Programa Nacional de Regionali-zacion Agraria 1979: Documento B, 93–105). In contrast to the preceding examples, low rates of out-circulation are found in fringe cantones such as Celica and Puyango in Loja, all of Bolivar, and Pangua in Cotopaxi. As noted earlier, these cantones differ considerably from highland Sierran ones, both culturally and in their minifundio situation.[18]

Statistical findings from logistic regressions (Table 6.4, zero-order and multiple variable analyses) complement these observa-tions on circulation. Cantones where the hacienda/huasipungo system was strong are still characterized by large farms; under these circumstances, out-circulation associated with nearby mini-fundios yields a positive relationship with SIZE. In such can-tones, latifundios are long-standing, whereas minifundios may not

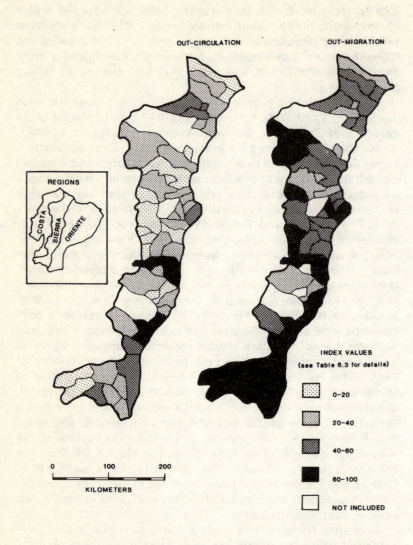

Figure 6.1 Population movements from the rural Sierra

be; but in other cantones, minifundio holdings that pre-date land reform are a principal feature. Hence, out-circulation and TIME relate directly. Cantones with characteristics indicating more traditional settings (where cultural ties would be stronger) have negative scores on MODERNITY, leading to an inverse relationship with out-circulation. Finally, that PRODUCTION ORIENTATION is not significant indicates out-circulation is equally likely in both semi-subsistence and market-oriented settings.

MIGRATION

Migration from the rural Sierra is more distinctly patterned than circulation (Figure 6.1). The southernmost province of Loja and the Oriente half of Azuay, just north, constitutes a zone of high out-migration. Another source is fringe cantones such as Chillanes, Chimbo, and San Miguel in Bolivar; Pangua in Cotopaxi; Santo Domingo in Pichincha; and others in Azuay and Loja. High out-migration also is found for cantones with urban agglomerations such as Ibarra in Imbabura, Banos in Tungurahua, Riobamba in Chimborazo, Loja in Loja, and Santo Domingo. By contrast, lesser out-migration levels characterize cantones centered on Ambato and Quito in the Sierra's northern half; and a less distinct but similar pattern is found in the vicinity of Cuenca.

Two generalizations are suggested. First, migration is more likely from areas that have a tradition of wage labor (or have experienced labor commodification in their transition from a feudal, hacienda-based economy). This characterizes both fringe cantones and urban agglomeration ones. Second, migration is more likely from remote cantones where local or nearby employment is limited and accessibility to other locales is poor, which reduces the practicality (viability) of circulation. But where conditions are reversed, circulation is an attractive alternative, particularly if ties to land and culture, discussed above, are strong. This second generalization accounts both for high rates of out-migration from Azuay and Loja and for lesser rates from cantones surrounding Ambato and Quito.

Statistical findings from logistic regression (Table 6.4, zero-order and multiple variable analyses) complement the preceding observations on migration. Fringe cantones, which include several in Azuay and Loja as well as those just mentioned, are typified by agricultural production in perennials, a domestic/export market orientation, and medium or small farms that are relatively recent in origin. These traits coincide with a strongly inverse relationship between out-migration and PRODUCTION ORIENTATION, and somewhat lesser but also inverse relationships with SIZE and TIME. Cantones with urban agglomerations have socioeconomic characteristics indicating more contemporary settings, which underlies the positive relationship between out-migration and MODERNITY. But the strength of that relationship is reduced because circumstances in Azuay and Loja, representing more traditional settings, impel their populations towards migration, while more contemporary cantones in the sphere of Ambato and Quito are prone towards circulation.

LAND REFORM AND MOVEMENT FROM THE RURAL SIERRA

These are linked in a number of ways, but in a less distinct manner than earlier accounts suggest. Legally voiding the hacienda as a socioeconomic form, together with economic conditions in the countryside, stimulated family strategies in which circulation and/or migration had an important role. Most affected were cantones with a strong hacienda tradition, which were concentrated in central and northern portions of the rural Sierra. In its southern highlands, several landholding forms had a strong presence; haciendas or related latifundio types, long-standing minifundios, and medium-size farms. On the eastern and western edges of the cordillera, both north and south, there also were areas settled through colonization from the 1950s onward, referred to here as *fringe* cantones. Different processes of out-circulation and out-migration characterize each of these locales and landholding types within them.

Further, land reform represented formalization of processes already ongoing; i.e., Ecuador's movement from feudal to commodified labor and production structures (Peek and Standing 1979, 1982a, 1982b). By 1964, the hacienda system was already on a road to transformation (Handelman 1980). In the north, where the system was more pervasive and land use more intensive, peasants had mobilized to claim rights which they believed were associated with the land rather than granted by the hacendado. In the south, where hacienda land was often under- or unutilized, squatter settlement and associated class conflicts were common. Under these circumstances, and the mounting realization that contemporary agriculture could yield greater profits, hacendados themselves (particularly in the north) took an active role in promoting land reform legislation and the changes it foreshadowed. Eventually, agriculture would become more capital-intensive, less dependent on labor, and urban agglomerations would increase in both size and importance. In this context land reform is but one aspect of a broader, multifaceted process; and migration and circulation are mechanisms by which cataclysmic shifts in society are smoothed.

Finally, it is relevant to briefly consider places to which these population movements are directed, another facet of the adjustment process just discussed. Circulation, except to Quito, is unfocussed, even if only a single origin canton (or province) is considered. But migration from the rural Sierra is noticeably directed: to proximate locales; to Quito and Ambato in the Sierra; to Santo Domingo, Quevedo, Guayaquil, and the Guayas Basin in the Costa; and to all Oriente provinces (Appendix 6.1). These destinations are similar to those found by Preston's examination of migration from five rural Sierran parroquias, an areal unit smaller than the canton (Preston 1980, Preston and Taveras 1976; Preston, Taveras, and Preston 1979). Economic growth and related opportunity is a common denominator among popular destinations, but given those attributes, more proximate places are often selected. Thus, northern Sierran migrants are more likely to choose the cantones of Santo Domingo or Quevedo in the Costa, or Napo province in the Oriente; while destinations of southern Sierran migrants are often Guayaquil, the Guayas Basin, or in the Oriente, the provinces of Morona Santiago and Zamora Chinchipe. Hence, proximity seems to operate in two ways: as a factor in its own right, and as a filter to choose among places offering similar levels of opportunity.

Broadening the perspective

This chapter examines individual movements from Ecuador's rural Sierra during the period 1974–82, giving particular attention to the operation of land reform policies. Instead of summarizing its empirical findings, discussion returns to three broad concerns.

First, migration and circulation within the same population were compared in order to identify differences in each process, a unique focus. Evidence here suggests that personal attributes operate similarly in both processes, albeit at different levels of intensity. With regard to place effects, circulation in Ecuador was associated with more traditional socioeconomic settings; migration with more contemporary ones. In this space-economy, then, circulation and migration are distinctly *different* processes; particularly in the role of *place characteristics* associated with development.

A second concern addressed by this chapter is drawing substantively informed conclusions from statistical analyses of readily available data with broad geographic coverage, of which censuses are an example. Illustrated throughout the book, but especially here, is that an essential element of the task is *knowledge of place* (or more broadly, knowledge of one's subject). Field experience is critical, even if the topic is one's own country. Also valuable are regional geography accounts, ethnography studies from anthropology, and special reports such as Programa Nacional de Regionalizacion Agraria (1979).

To elaborate, one might be satisfied to differentiate circulation and migration in terms of, say, a traditional–contemporary distinction. But, by using qualitative knowledge of Ecuador to interpret statistical findings from census data, our understanding of those processes is considerably more detailed.

Qualitative knowledge pertained to Ecuador's ongoing transition from feudal to commodified labor and production structures, its land reform policies which were a component of the transition, how policy articulation varied in different rural settings, and details on cantones themselves at ground level. For analytical purposes, this was complemented by statistically derived indices which distinguished cantones on the basis of recency of landholdings, production for semi-subsistence or market purposes, size of landholdings, and modernity of their socioeconomic environment.

Qualitative comprehension, or sense, of Ecuador as a place was a critical ingredient of the analyses. Without it, or drawing on statistical findings alone, one might conclude, for example, that out-migrants originated from small farms and out-circulators from larger farms. In fact, both originated from small farms, but under different socioeconomic conditions at the local level. Circulation reflects identification with a cultural or ethnic community and holding land (usually a small parcel); it may be a response to recent circumstances or a long-standing community tradition; and it may involve artisannal activity, urban enterprise, or agricultural employment. Migration is associated with contemporary (rather than traditional) production systems; opportunities to settle new lands (e.g., in the Oriente); earlier migration experiences (often through land settlement or colonization); and giving in to or accepting the transition to labor commodification. Playing a role in both circulation and migration were limited land, land deterioration, and other aspects of population pressure; conflicts between social classes and/or communities; and natural disasters such as the 1968 drought in southern portions of the Sierra. Land reform, in legally voiding the bond between hacendados and huasipungueros, created economic pressures that fomented both circulation and migration; it also provided mechanisms by which settlers of unoccupied lands, as well as former huasipungueros, could gain title. But land reform itself represented formalization of Ecuador's ongoing movement from feudal to commodified labor and production structures, which had begun some years before the 1964 legislation.

Comprehending the socioeconomic *heterogeneity* of the rural Sierra, both as a place and in terms of processes underlying population mobility, is important in its own right. But such knowledge also serves as a counter to social science generalizations that are overly facile.

A third issue is understanding *regional change*, a question of central concern to the study of development by geographers and regional scientists. Land reform policies and associated events had dissimilar effects and promoted different responses among individuals and entrepreneurial endeavors, depending on one's personal, economic, and locational characteristics. This impacts on, and extrapolates into, regional change because, in Ecuador and elsewhere, distinct sets of characteristics dominate different locales to create areal complexes. Consider, for example, the divergent profiles of Sierra and Costa in both their urban and rural sectors; and within the rural Sierra, canton variations in agrarian and socioeconomic structure, described earlier in the chapter. Hence, *apparently aspatial* policies, such as land reform, have highly variable place-to-place effects. By similar reasoning, *explicitly spatial* policies also exhibit spatial impacts that differ from those anticipated. An example is Venezuela's policies to spatially decentralize both educational opportunity and economic activity, discussed in Chapter Five.

More generally, examining land reform and population mobility is a step towards establishing that regional change, or intranational variations in development, represent the local articulation of world economic and political events, donor-nation actions, and policies of Third World governments. Central to this perspective is treating such forces and related elements in their *own right*, rather than as examples of either orthodox or political economy conceptualizations. The latter portrayals of development have dominated our vision for some time and led to considerable gains in knowledge. But even if seen as complementary rather than competing explanations, conventional portrayals divert attention from *tangible* forces playing a major role in today's Third World.

Hence, this study disregards dependency aspects of Ecuador's patron/hacienda system; neoclassical economic aspects of utility-maximizing behavior by peasants; and the presence of both elements in economic motivations underlying land reform policies. Deemed more important was delineating the operation of land reform in population movements from rural Sierra locales, giving particular attention to the mediating role of place variations in agrarian and socioeconomic structures.

There is another way to view the research strategy towards which this chapter, and the book, is pointing. Highlighting similarities between places has been important, and conventional development paradigms are a segment of that task. But a more current need is to focus on place differences in order to gain a better understanding of local variation and its role in Third World development (or change) at all spatial scales. Ultimately, generalizations should emerge, and that remains as an objective – but generalizations which are rich in detail, recognize the *heterogeneity* of development processes, and emphasize *visible* mechanisms underlying regional change.

This strategy has an additional and more immediate consequence. Knowledge concerning the local, or even national, articulation of forces that play a fundamental role in shaping Third World development is lacking. Examples of such forces include import substitution industrialization policies, International Monetary Fund (IMF) loan practices, land reform policies, or the operation of major commodity markets such as that for petroleum. These are major determinants of place-to-place differences and regional change trajectories at all spatial scales, and further study of such forces in their own right should be high on our research agendas. In this spirit, Chapter Seven considers several policies that are elements of Ecuador's import substitution industrialization program, their likely impact on individuals and economic enterprises, and the way the interaction of these forces translates into regional change differentials.

Appendix 6.1 Provinces, cantones, and selected urban areas of Ecuador

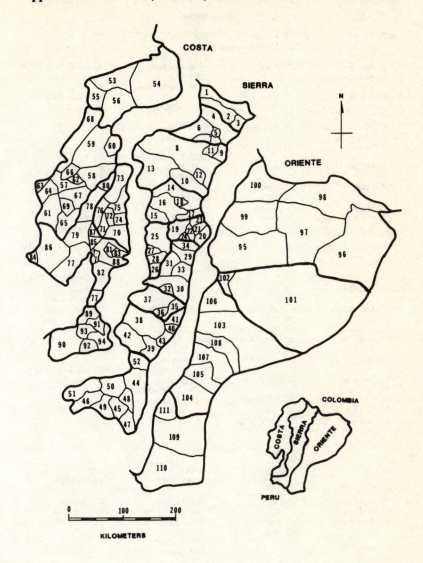

Appendix 6.1, continued

Sierra Region

CARCHI

**1. Tulcan
2. Espejo
3. Montufar

IMBABURA

**4. Ibarra
5. Antonio Ante
6. Cotacachi
7. Otavalo

PICHINCHA

**8. Quito
9. Cayambe
10. Mejia
11. Pedro Moncayo

12. Ruminahui
**13. Santo Domingo

COTOPAXI

14. Latacunga
15. Pangua
16. Pujili
17. Salcedo
18. Saquisili

TUNGURAHUA

**19. Ambato
**20. Banos
21. Patate
22. Pelileo
23. Pillaro
24. Quero

BOLIVAR

25. Guaranda
26. Chillanes
27. Chimbo
28. San Miguel

CHIMBORAZO

**29. Riobamba
30. Alausi
31. Colta
32. Chunchi
33. Guamote
34. Guano

CANAR

35. Azogues
36. Biblian
37. Canar

AZUAY

**38. Cuenca
39. Giron
40. Gualaceo
41. Paute
42. Santa Isabel
43. Sigsig

LOJA

**44. Loja
45. Calvas
46. Celica
47. Espindola
48. Gonzanama
49. Macara
50. Paltas
51. Puyango
52. Saraguro

Costa Region

ESMERALDAS

**53. Esmeraldas
54. Eloy Alfaro
55. Muisne
56. Quininde

MANABI

**57. Portoviejo
58. Bolivar
59. Chone
60. El Carmen
61. Jipijapa

62. Junin
**63. Manta
64. Montecristi
65. Pajan
66. Rocafuerte
67. Santa Ana
68. Sucre
69. 24 de Mayo

LOS RIOS

70. Babahoyo
71. Baba
72. Puebloviejo

**73. Quevedo
74. Urdaneta
75. Ventanas
76. Vinces

GUAYAS

**77. Guayaquil
78. Balzar
79. Daule
80. El Empalme
81. Milagro
82. Naranjal
83. Naranjito

84. Salinas
85. Samborondon
86. Santa Elena
87. Urbina Jado
88. Yaguachi

EL ORO

**89. Machala
90. Arenillas
91. Pasaje
92. Pinas
93. Santa Rosa
94. Zaruma

Oriente Region

NAPO

**95. Tena
96. Aguarico
97. Orellana
98. Putumayo

99. Quijos
100. Sucumbios

PASTAZA

101. Pastaza
102. Mera

**MORONA
SANTIAGO**

103. Morona
104. Gualaquiza
105. Limon Indanza
106. Palora
107. Santiago
108. Sucua

**ZAMORA
CHINCHIPE**

109. Zamora
110. Chinchipe
111. Yacuambi

** Indicates canton contains an urban area referenced in text

7 Policy aspects of development and regional change II: The juxtaposition of national policies and local socioeconomic structures in Ecuador[1]

World economic and political events, donor-nation actions, and national policies affect local areas both directly, and indirectly by inducing spatial variations in individual or entrepreneurial actions. Indirect effects have taken precedence in earlier chapters which examined population movements in Costa Rica, Ecuador, and Venezuela and labor market experiences in Venezuela. This culminated in documenting a chain from national policy, through place characteristics associated with development, which act as a mediating agent, to individual and entrepreneurial behavior.

Direct effects on regional change also have been evidenced, but as an ancillary finding. Hence, in Chapter Five we saw that Venezuela's socioeconomic landscape was materially affected by its policies to spatially decentralize both educational opportunity and economic activity; and in Chapter Six, that land reform and related forces in Ecuador primarily impacted cantones with large landholdings, and more so in the Sierra than in the Costa.

In this chapter, focus shifts directly to *regional change*, examined in terms of several forces. After background observations concerning the treatment of policy by geography and regional science, the chapter identifies patterns of change among Ecuadorian cantones between 1974 and 1982, a period of dramatic economic shifts. Attention then turns to selected policies that influenced this change; agricultural credit, agricultural pricing, and monetary exchange rate measures which are elements of Ecuador's import substitution industrialization program. Finally, canton variables mediating, channeling, or determining impacts of these policies are employed in a discriminant analysis to distinguish among canton groups, thus establishing a direct link between policy measures and regional change.

Background

Seeing national policy, world events, and donor-nation actions as critical elements in Third World development is not new in itself, but stressing their local articulation represents a departure from the usual. In a spatial frame of reference, emphasis has been on broad tendencies such as urban bias or polarization (Lentnek 1980; Lipton 1976; Renaud 1981; Richardson 1981; Todaro and Stilkind 1981), rather than on local effects. And when effects are more finely differentiated, focus tends to be on economic sectors rather than places; for example, trade, fiscal, and monetary policy biases that favor manufacturing over agriculture (Adams, Graham, von Pischke 1984; Lipton 1976; Meier 1984).

Since places vary in economic structure, national policies (and other exogenous forces highlighted in this book) *must* be a prominent factor in the spatial differentiation of regional change and uneven growth/decline within nations. Nevertheless, writings concerned with the spatial form of development and policy tend to view this conjunction as a mechanistic occurrence related to urban system characteristics, city size, accessibility, and spatial interactions between urban areas and their hinterlands or among urban areas themselves. These elements are seen to induce spatial variation in development by localizing change, through economies of scale or 'circular and cumulative causation', and/or by transmitting it differentially among places. Other place characteristics and their role in creating, channeling, distorting, or frustrating policy impacts and development tend to be assumed away, or surrogated under size and accessibility (Richardson 1981).

Hence, attention has been diverted from place attributes (as well as other forces) that underlie the correlation between, say, city size and development differentials within a nation, and *spatial fetishism* typifies policy-oriented writings (Gore 1984; Sheppard 1982). Early work advocated a centralization approach whereby investments, policy initiatives, administrative energies, and the like are concentrated in large or primate cities, from where they should trickle down to other places (Hirschman 1958). Disappointing performance under this strategy led to rural-oriented, bottom-up approaches stressing dispersal of policy efforts to counteract (previous) urban biases (Lipton 1976; Rondinelli and Evans 1983; Rondinelli and Ruddle 1978; Stohr and Taylor 1981). More recently, intermediate cities have been seen as ideal agents of change in regional economies (Rondinelli 1983).

In a fundamental sense, then, policy approaches concerned with spatial form have been *place-blind*, an irony given that geographers and allied professionals have been major players in elaborating the focus. This body of literature tends to reflect paradigmatic perspectives of an orthodox nature. Research from a political economy perspective has been more sensitive to place, often identifying factors underlying conditions such as primacy or urban bias (e.g., Taylor 1980), but policy prescriptions are rarely the goal. There also has been some bridging of postures; Lipton (1976) and Todaro and Stilkind (1981) are examples.

Here, the issue of understanding and narrating uneven regional change is addressed in the following manner. First, the substantial role of national policies, which all governments employ to direct their economy, is acknowledged.

Second, policies are formulated and channeled according to resource scarcities, institutional rigidities, political constituencies, and the like. But aside from distortions related to these parameters, the impacts of state intervention are not uniform across the space-economy. Rather, they are *conditioned* by *local socioeconomic structures* and vary accordingly. For example, if policy impacts depend on whether farmers produce for subsistence, domestic, or export markets, places with a preponderance of one such group will be affected differently from others. This is relevant because, as noted in Chapter Six, elements of regional structure (persons, households, entrepreneurial entities, urban agglomerations, infrastructures, resources, production possibilities, and the like) are located so as to form areal complexes that are (more or less) distinct in socioeconomic characteristics. Hence, an apparently *aspatial policy*, such as land reform, may lead to marked *differences* in *regional change*. This occurs through direct effects on structural elements and indirectly by inducing spatial differentiation in individual or entrepreneurial actions.

Third, to appropriately account for uneven regional change, analytical frameworks need to consider the juxtaposition of policy and other external forces with place characteristics. Attributes emphasized in earlier studies, such as accessibility and city size, are acknowledged. But this chapter concentrates on characteristics that differentiate places at a more detailed, fine-grained, local level which, as noted throughout the book, have been neglected.

Regional change in Ecuador, 1974–82

The basis for identifying shifts in Ecuador's space-economy are twenty-seven canton-level variables representing conditions in

1974 and, for the period 1974–82, absolute and relative changes in economic and human resource attributes. These were subjected to principal components analysis, yielding four dimensions (Table 7.1) and associated component scores (Appendix 7.1). The latter were then employed to group cantones according to their regional change experiences. Interpretation of principal components and canton groupings draws on the preceding chapter's discussion of Ecuador and underlying materials such as Programa Nacional de Regionalizacion Agraria (1979), a highly detailed account at the canton level of agrarian structure and its historical antecedents. Canton locations are found in Appendix 6.1.

Principal components

The first principal component portrays *absolute employment change*. Positive scores on this dimension are recorded by larger and more urbanized cantones that experienced absolute employment increases in manufacturing, commerce, and service sectors of the economy and in professional, clerical, and vendor occupations; the proportion of highly educated persons also increased. Places exemplary of this include Quito (+6.12) in Pichincha province, Guayaquil (+7.39) in Guayas, Cuenca (+1.16) in Azuay, Ambato (+1.00) in Tungurahua, and Machala (+0.92) in El Oro; Ecuador's five largest urban areas. At the opposite extreme are rural, sparsely populated, often remote cantones experiencing little absolute change; for example, Muisne (-0.60) in the Costa province Esmeraldas and, in the Sierra, Celica (-0.65) in Loja and Paute (-0.76) in Azuay.

The second principal component portrays *economic structure shifts*. Positive scores on this dimension are recorded by cantones experiencing large relative increases in total population; in agriculture-mining, manufacturing, commerce, and transportation sector employment; and in persons with vendor, clerical, and agriculture-mining occupations. These cantones also recorded absolute increases in agriculture-mining employment. Prime examples are Oriente cantones associated with agricultural colonization such as Morona (+2.29) and Gualaquiza (+1.19) in Morona Santiago province and in Napo, where petroleum development also occurred, Orellana (+5.10) and Putumayo (+6.27). Negative scores are found for cantones with agrarian structures that are stable or in slow transformation. Examples include a number of

Table 7.1 Principal components analysis of socioeconomic variables representing change among Ecuador cantones for 1974–82, varimax rotation

Variable	Mean	I	II	III	IV	Communality
			Component			
Percent change in						
Total population	17.66	0.12	0.87	0.06	-0.02	0.78
Agriculture-mining sector employment	-5.77	-0.08	0.83	0.30	-0.28	0.87
Manufacturing sector employment	19.08	0.17	0.64	-0.33	0.21	0.59
Commerce sector employment	38.06	0.05	0.85	-0.16	0.06	0.76
Transportation sector employment	141.56	-0.14	0.44	-0.27	0.06	0.29
Service sector employment	109.57	-0.10	0.32	-0.66	-0.09	0.55
Persons seeking first work	193.18	0.04	0.09	-0.69	-0.05	0.49
Persons with vendor or clerical occupations	34.65	0.04	0.89	-0.02	0.05	0.79
Persons with farming occupations	-6.13	-0.08	0.80	0.36	-0.31	0.87
Absolute change in						
Agriculture-mining sector employment	-917.45	-0.09	0.54	0.58	-0.27	0.70
Manufacturing sector employment	549.89	0.95	0.06	0.01	-0.08	0.91
Commerce sector employment	764.13	0.96	0.05	0.01	-0.12	0.94
Service sector employment	2051.94	0.98	0.03	-0.01	-0.06	0.96
Persons with professional occupations	882.70	0.96	0.08	0.06	-0.09	0.94
Persons with vendor or clerical occupations	526.16	0.97	0.06	0.01	-0.10	0.95
Persons with farming occupations	-957.99	-0.07	0.47	0.68	-0.31	0.78
Dependency ratio ((population<15 + population>65)/population aged 15-65), 1974	1.00	-0.38	-0.36	-0.31	-0.20	0.41
Total population, 1974	58701.01	0.97	-0.04	-0.03	-0.08	0.96
Percent urban population, 1974	21.77	0.68	-0.07	0.12	0.41	0.65
Mean years of education of 1974–82 out-migrants	1.92	0.10	-0.02	0.64	0.14	0.45
Mean years of education of 1974–82 in-migrants	1.93	0.08	0.02	0.49	0.09	0.26
Mean occupational status of 1974–82 out-migrants	34.68	0.06	0.11	0.45	0.26	0.57
Mean occupational status of 1974–82 in-migrants	37.81	-0.14	-0.39	0.30	-0.17	0.29
Percent in-migrants minus percent out-migrants among the economically active	3.11	-0.18	0.06	0.11	0.68	0.51
Percent in-migrants minus percent out-migrants, male population only	0.33	-0.17	0.06	-0.01	0.79	0.66
Change in percent of population with college education	1.30	0.66	-0.11	0.09	0.46	0.67
Change in percent of population with no education	-8.54	0.28	-0.14	0.11	0.45	0.31
Percent variance explained – by each component		27.1	19.7	11.4	8.1	
– cumulative		27.1	46.8	58.2	66.3	

n = 109

cantones in Manabi province, characterized by large cattle operations and/or coffee production on small- and medium-size holdings; e.g., Santa Ana (-1.01), Sucre (-1.26), and 24 de Mayo (-1.22).

The third principal component pertains to *production system reorientation*, either from traditional systems to wage labor and urban-focussed activities (+), or within commercial agriculture, from one crop/production system to another (−). Positive scores are recorded primarily by Sierran cantones where the educational level and occupational status of in- and out-migrants increased, as did the number of persons employed in agriculture. The latter is partially related to increased circulation brought about by land reform, as in Chimbo (+1.07) and San Miguel (+1.12) in Bolivar province, Guano (+1.36) in Chimborazo, and Azogues (+1.47) and Biblian (+1.35) in Canar. These latter two cantones also have built up artisannal activity in jewelry and pottery, as have Antonio Ante (+1.56) and Otavalo (+1.20) (Imbabura) in textiles. Other cantones shifted agricultural production to better reflect proximate urban markets (emphasizing cheese, flowers, fruits, and vegetables for example) and/or their population took up employment therein; this is found in Tulcan (+1.18) in Carchi, Quito (+1.06) in Pichincha, and Ambato (+1.04) and Pillaro (+1.07) in Tungurahua. The broad process represented by these examples is a transformation of post-land reform minifundio economies to ones addressing current conditions.

Negative scores on the *production system reorientation* dimension are recorded primarily by Costa cantones experiencing a proportional increase in new entrants to the labor force and in service and manufacturing employment. These cantones suffered severe disruption to their economic system − through blight-related curtailment of banana production in the early 1960s; land reform accompanied by hostilities among social classes, largely in the early and mid-1970s; and economic recession in the late 1970s, early 1980s. As a result, large landholdings were subdivided, cropping became more diversified, temporary and wage labor more prevalent, and the socioeconomic system was in continual flux. Examples from cantones that had considerable commitment to banana production include Santo Domingo (-1.15) in Pichincha; Babahoyo (-1.07) and Ventanas (-1.19) in Los Rios; Balzar (-1.29), Daule (-2.47), El Empalme (-2.29), Samborondon (-1.88), and Yaguachi (-1.61) in Guayas. Except for less land appropriation, similar changes occurred in cantones having large holdings devoted to cattle such as Bolivar (-1.63), Chone (-1.32), Jipijapa (-3.59) and Sucre (-1.76) in Manabi; and in cantones characterized by medium-size coffee fincas such as Pajan (-2.41), Rocafuerte (-1.33), Santa Ana (-2.23), and 24 de Mayo (-2.48) in Manabi.

The fourth dimension of regional change portrays *labor force shifts*, distinguishing cantones experiencing net in-migration (+) from those experiencing net out-migration (−) of the economically active, particularly male, population. Positive scores on this dimension are found for areas of frontier development, often proximate to the Peruvian border; for example, Calvas (+1.65), Celica (+1.23), and Macara (+2.24) in Loja province; Arenillas (+1.62) in El Oro, where smuggling from Peru also is rampant; and Banos (+1.56) in Tungurahua. Other areas with positive scores include Costa cantones characterized by relatively late land reform and labor-intensive primary activity such as rice cultivation and fishing; e.g., Yaguachi (+1.21) in Guayas, Pasaje (+2.18) and Santa Rosa (+2.23) in El Oro. Finally, positive labor force shifts are indicated for cantones with rapidly growing secondary cities, often themselves in frontier or commercial agriculture/fishing areas: Tulcan (+1.16) in Carchi, Ambato (+0.99) in Tungurahua, Riobamba (+1.79) in Chimborazo, Loja (+2.46) in Loja, Portoviejo (+1.41) in Manabi, Manta (+1.12) in Manabi, and Machala (+2.12) in El Oro. Negative scores indicating net out-migration of the economically active, particularly male segment of the labor force are found for large urban areas such as Quito (-1.27) and Guayaquil (-1.29), for traditional areas throughout the Sierra, and for Costa cantones with stagnating economies and/or overpopulation, particularly in the provinces of Los Rios and Esmeraldas.

Canton groupings

To identify regional change profiles for the period 1974–82, Casetti's (1964, 1965) discriminant iterations grouping algorithm was applied to component scores. This provides a fourfold classification of cantones as (1) Major Urban Entities, (2) Regional Centers, (3) Commercial Agricultural Areas, and (4) Traditional Agricultural Areas (Appendix 7.2). For each group, average component scores are shown in Table 7.2 and their spatial distribution in Figure 7.1.

MAJOR URBAN entities include Ecuador's two largest cities, Quito and Guayaquil, where population, political power, wealth,

Table 7.2 Average principal component scores for canton groups, indicating the nature of socioeconomic change from 1974 through 1982

Canton group	Absolute employ- ment change (I)	Average component score Economic struc- ture shifts (II)	Production system reorientation (III)	Labor force shifts (IV)
Major urban	6.76	0.26	0.35	-1.28
Regional center	0.03	0.33	0.43	1.04
Commercial agriculture	0.01	-0.30	-1.56	0.03
Domestic agriculture	-0.32	-0.12	0.37	-0.72

and urban-based economic activity are concentrated. These areas exhibit high levels of growth in absolute employment (Component I, +) and net out-migration among the economically active, particularly male segment of the labor force (Component IV, −). The latter reflects the following conditions. Like primate cities in other nations, Quito and Guayaquil have been major focal points for migration, but growth in job opportunities lagged behind the influx of people, leading to overpopulation and under- or unemployment. But in the 1970s, Ecuador's economic advances in petroleum and commercial agriculture created employment opportunities throughout the country, especially for males; and together with social change related to land reform, this led to labor force movements away from these major urban places.

REGIONAL CENTER cantones include Ecuador's third largest city, Cuenca; most of its intermediate-size cities; and several Oriente areas associated with petroleum exploitation. These places moved towards more robust economies by undergoing favorable shifts in economic structure (Component II, +) and, among Sierra cantones, by reorienting production systems to better conform with current conditions (Component III, +). For these reasons, cantones in this group also experienced high levels of net in-migration by the economically active (Component IV, +).

COMMERCIAL AGRICULTURE cantones, located entirely in the Costa, experienced reorientation in economic activity from one crop/production system to another (Component III, −). This largely involved a decreasing commitment to bananas while diver-

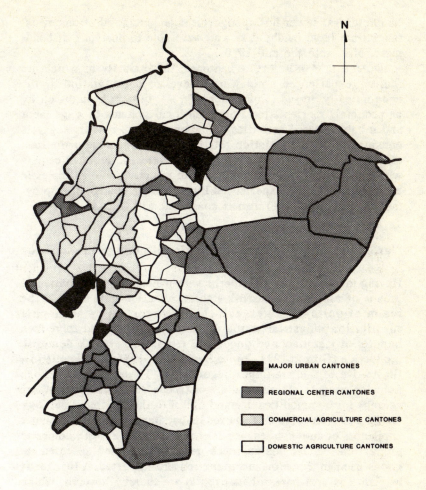

MAJOR URBAN CANTONES

REGIONAL CENTER CANTONES

COMMERCIAL AGRICULTURE CANTONES

DOMESTIC AGRICULTURE CANTONES

Figure 7.1 Spatial distribution of canton groups in Ecuador

sifying into other croppings; an increasing dominance of medium-size landholdings (100-500 hectares) as larger ones were disbanded; and some colonization. At the same time, this group experienced relatively less change in economic structure (Component II, −) because commercial agriculture remained the dominant economic activity and land reform was less effective than in the Sierra. The latter was due to an established segment of medium-size landholdings, fewer huasipungo arrangements, and strong resistance from large landholders who were able to postpone disbandment, often into the mid-1970s.

DOMESTIC AGRICULTURE cantones, the majority of which are Sierran, experienced a noteworthy level of out-migration by the economically active (Component IV, −), in part brought on by an economic reorientation from traditional systems to wage labor and urban-focussed activities (Component III, +). This suggests considerable overpopulation relative to economic opportunities, even after reorientation; an interesting contrast to regional centers where reorientation was linked with in-migration. That domestic agriculture cantones experienced little absolute change in employment (Component I, −) also is consistent with this scenario.

National policies influencing regional change in Ecuador

Having identified growth patterns within Ecuador's economy, and groups of cantones representing them, attention now turns to the realm of policy. In general, this is governed by an import substitution industrialization (ISI) strategy intended to reduce imports of manufactured goods and replace them with domestic production (Gibson 1971; World Bank 1979, 1984; Zuvekas 1975). Illustrative of this are import regulations. Throughout the 1970s, Ecuador tried to maintain minimal tariffs and unlimited import ceilings for commodities deemed useful to the economy, but high tariffs and restrictive ceilings for luxury items, such as consumer appliances or automobiles, and for goods competing with domestic production. Under this approach, raw materials and intermediate goods used in Ecuadorian production are imported at low tariff levels. The cost bases of its exports are thereby lowered, which enhances their competitiveness in world markets and increases inflows of needed foreign exchange. At the same time, Ecuadorian goods for domestic markets are protected from foreign competition by tariffing the latter at high levels. Finally, when foreign-produced items are purchased anyway, their high tariff structure channels much of the outlay to Ecuador's government, in theory for use in programs that benefit development. During the 1970s, this process was enhanced by, but also became dependent on, massive inflows of foreign exchange from petroleum exports. Another aspect of ISI is its impact on the relative strength of economic sectors. Agricultural pricing, credit, and subsidy practices favored commercial over domestic production; they also maintained urban food prices at minimal levels, thereby overtaxing agriculture relative to other economic sectors.

The net effect of ISI policies, in Ecuador as elsewhere, has been an urban/industrial/modern sector bias in development (Lipton 1976; Todaro and Stilkind 1981). In spite of this, agriculture maintained its position as an essential element of Ecuador's economy (World Bank 1984: 77). It declined from somewhat more than 73.0 percent of total exports in 1970 to 23.0 percent in 1983 (World Bank 1984: 128), but given the policy milieu and petroleum's overwhelming performance (e.g., 74.0 percent of 1983 exports), maintaining that share is impressive. Agriculture also is the most pervasive sector, directly or indirectly employing an exceptionally high percentage of the population and of the national territory.

Agriculture and related activities thus remain at the heart of Ecuador's economy. For this reason, agriculture is used here as the primary sector through which relationships between policy and uneven development are more fully explored. In so doing, detailed attention is given to agricultural pricing, agricultural credit, and monetary exchange rate policies.

Price policies

Ecuador guarantees a maximum return for agriculturalists and minimum cost for consumers on selected food products. The former is implemented through Empresa Nacional de Almacenamiento y Comercializacion (ENAC); the latter through Empresa Nacional de Productos Vitales (ENPROVIT).

ENAC buys crops at officially set prices and sells at somewhat higher levels to cover storage and handling expenditures, targeting both domestic consumption and export commodities. Objectives of this program include stabilizing farm income and increasing production through the incentive of an adequate and guaranteed return (Thirsk 1976; Tolley, Thomas, and Wong 1982; Zuvekas 1975), but these goals have been unrealized for a number of reasons.

First, ENAC prices have not consistently matched production costs, particularly for domestically sold commodities. In part this occurs because domestic crop support levels are determined by a legally prescribed framework, whereas export crop supports are determined from international prices. Also, domestic crop prices are reviewed and adjusted infrequently, thus lagging in their reflection of seasonal or longer-term supply/demand shifts (Economic Perspectives 1985: 51–7; World Bank 1984: 78–81).

Second, ENAC practices often have amplified, rather than reduced, supply/demand imbalances (Economic Perspectives 1985). In 1981, for example, a record hard corn harvest was recorded, in part because of ENAC's support price; the following year's harvest was even larger. Initially, ENAC purchased and stored sufficient quantities to maintain market order, but by the second year its storage facilities were filled. As a result, surpluses went directly into local markets, which drove prices well below the guaranteed level. ENAC responded by selling abroad at a considerable loss in order to reestablish its storage capability. In 1983, therefore, ENAC had insufficient stock to cover devastation related to El Nino floods, which severely reduced corn harvests and destroyed corn supplies.

Third, ENAC's purchasing practices have been a destabilizing factor by targeting contemporary, large-scale agriculturalists more effectively than minifundistas. This reflects the farmer's ability to comprehend information concerning support prices, but it also appears that information is not uniformly available. Further, ENAC outlets tend to be located in larger, more formally organized markets, whereas many small-scale farmers rely on informal sector middlemen, in part because of their accessibility. Hence, whether by accident or design, ENAC often buys from larger operations, as Franklin and Penn (1985: 5) have found for rice.

These practices are not explicitly spatial, but they operate in that manner because locales differ considerably in agricultural characteristics. Destabilization related to ENAC's treatment of corn, for example, especially impacted Manabi, Los Rios, and Guayas provinces in the Costa, where 60 percent of Ecuador's production takes place (Instituto Nacional de Estadistica y Censos 1983). That ENAC's purchasing practices disadvantage minifundios and other domestic market producers, resulting in poor support for crops such as potatoes and kidney beans, primarily affects the Sierra. At the same time, ENAC support programs bolster Costa locales with larger-scale export-oriented operations, such as Los Rios, where soy bean production is concentrated (Economic Perspectives 1985: 69–77).

Turning now to ENPROVIT, the program providing food to consumers, its outlets carry hundreds of products including rice, meat, legumes, bread, flour, butter, corn, potatoes, sugar, milk, coffee, soybeans, cotton lint, and cotton seed. Prices for these are well below free market levels; in the early 1970s, often more than 50 percent so (Economic Perspectives 1985: 26; Lawson 1986: Table 5). Lesser-income consumers benefit and continuation of low urban wages is facilitated, which aids import substitution industrialization. But domestic market producers suffer in that ENAC support for them is least effective; yet selling elsewhere means competing against ENPROVIT's artificially low prices. More generally, the combination of ENAC/ENPROVIT practices is part of ISI's implicit taxation policy whereby resources are transferred from rural, (more) traditional sectors to urban, (more) contemporary ones, ultimately leading to stagnation in the former (Thirsk 1976; Zuvekas 1975).

Evidence of stagnation, and its locational specificity, is provided by potato production. This occurs almost exclusively in the Sierra, 24 percent in Chimborazo province alone (Simmons and Ramos 1985). Area planted to potatoes declined from 47,220 hectares in 1970 to 30,380 in 1980, a 35.7 percent reduction; and the decline in total output, 40.3 percent, was even greater (Lawson 1986: Table 6). Similar cutbacks are found for barley and kidney beans (Lawson 1986: Table 6), other major products of the Sierra.

Credit policies

During the 1970s, agriculture employed approximately 48 percent of the labor force, contributed up to 76 percent of Ecuador's exports, and averaged 18.5 percent of the gross domestic product. It received, however, only 14.7 percent of total credit (Ramos 1984: Table 7).

Within agriculture, Banco Nacional de Fomento (BNF) is the most prominent lending institution, providing approximately 40 percent of the loans from 1970 through 1982 (World Bank 1984: 82–3). As a government-supported entity, BNF's constitution mandates prioritizing credit to small-scale farmers (less than five hectares). Nevertheless, this approximately 77 percent of the agricultural sector received, on average, only 28 percent of BNF loans; the remainder primarily went to medium and large-scale agriculturalists (Ramos 1984).

This bias results in part from interest rate regulations. The legal interest ceiling is 12 percent, but during the 1970s, inflation ranged from 8 to 23 percent. Hence, the real rate of interest was often less than zero, thus reducing (or removing) saving incentives, shrinking credit availability, but cheapening that credit which is available. This encouraged political patronage in favor of larger agriculturalists.[2] As Ramos (1984: 47) observes,

> Interest rates kept low by government policy have tended to have effects that discriminate against small farmers. Due to a lack of large amounts of resources, cheap credit must be rationed. The procedures usually are politically determined and provide opportunities for corruption, cronyism, and favoritism.

Small farmers, many of whom are illiterate, also are disadvantaged by complex and costly borrowing procedures. BNF requires extensive documentation to obtain a loan, including an identification card, voting certificate, proof of income tax payment, and production report for the loan period. In addition, minifundistas often distrust written procedures that have little meaning in more traditional cultures. But the greatest impediment to credit is proof of land ownership, which automatically excludes renters and those without title (Ramos 1984).

Given that small farms and traditional agriculturalists are concentrated in the Sierra, credit disbursement has distinct spatial biases. Although the Sierra contains 62.1 percent of all farm units and 38.7 percent of all farm hectarage, it averaged only 34.1 percent of BNF credit (and much of that went to larger units). By contrast, the Costa contains 33.1 percent of all farm units, 47.3 percent of the hectarage, but received 60.2 percent of BNF credit (Lawson 1986: Table 9). As an offshoot of this disparity, the cost of borrowing also varies regionally, disadvantaging the most needy areas. This occurs because credit must be obtained elsewhere if not through BNF, and non-institutional channels such as money lenders charge considerably higher rates, averaging more than 50 percent (!) in 1982 (World Bank 1984: 82).

Monetary exchange rate policies

Throughout the 1970s Ecuador pursued a fixed exchange rate policy, maintaining the sucre's value higher than its free market level (World Bank 1984: 10–12).[3] This policy had two adverse effects. First, it artificially cheapened imports, thus undercutting domestic producers of similar goods. Second, an inflated sucre made Ecuador's exports more expensive, and hence less competitive, in world markets.[4] But there also were gains through overvaluation. First, among the artificially cheapened imports are raw materials and inputs of intermediate goods to manufacturing. By reducing production costs, while selectively employing high tariffs to stem competition from imports, industrial expansion, or ISI, was facilitated. Second, overvaluation increased petroleum exports because of high demand elasticity in world markets and Ecuador's comparative advantage in that commodity. This enabled producers to recoup losses associated with overvaluation by selling larger quantities, while providing a massive inflow of foreign exchange for other development purposes.

Concerning agriculture, two points are relevant. First, demand for, and production of, domestic foodstuffs declined precipitously because overvaluation and abundant foreign exchange lowered the cost of comparable imports. Depending on the food category, imports grew at average annual rates of 20 to 54 percent in the 1970s (Lawson 1986: Table 2) while domestic food production, already penalized by ISI policies, suffered. Among basic dietary ingredients, potato output declined 40 percent from 1970 through 1980, barley declined 69 percent, and kidney beans 36 percent (Lawson 1986: Table 3). Also, that yields remained approximately constant indicates domestic production systems were virtually unchanged, a form of stagnation.

A second point is overvaluation's uneven effect within agriculture. Machinery, chemicals, fertilizers, seeds, and other inputs to capital-intensive production became relatively less expensive. Upward pressures on wages also were reduced in that imports lowered food costs. Hence, although commercially-oriented agriculture experienced deleterious effects from overvaluation, by being less competitive in world markets for example, it benefited by an implicit subsidization of capital-intensive production systems (Cleaver 1985; Economic Perspectives 1985). More traditional agriculturalists, with production systems relying largely on family labor, gained little in this regard. Within agriculture, then, more contemporary entities were buffeted from the negative impacts of Ecuador's exchange rate policies, while lower-income, traditional producers bore only negative consequences.

Spatial outcomes of exchange rate practices include the following. Agricultural areas producing for domestic markets under more traditional systems were most disadvantaged; these are largely located in, and dominate, the Sierra. Areas producing for domestic markets, but employing more contemporary production systems, were somewhat less disadvantaged; these are largely in the Costa. Areas producing for export markets under more contemporary production systems, also located primarily in the Costa, experienced mixed effects. Economies with components related to petroleum production benefited; these are primarily in northern portions of the Oriente and Costa. Finally, urban areas, throughout the Sierra and Costa, tended to benefit from exchange rate policies, but for most, the economic character of their rural hinterland was a strong modifying element.

National policies and local growth outcomes

The preceding section established that national policies pertaining to commodity prices, agricultural credit, and monetary exchange rates had spatially differentiated impacts on Ecuador's economy. Building on that analysis, attention now turns to linking national, or macroeconomic, policies; local structural conditions that mediate, or determine, policy impacts; and regional change in Ecuador from 1974 through 1982 (Figure 7.2). The first step is to identify variables representing pertinent *structural* conditions at the canton level. Second, these structural variables are employed to statistically discriminate between Regional Center, Commercial Agriculture, and Domestic Agriculture canton groups, each of which represents different regional change experiences for the period 1974 through 1982 (Appendix 7.2). The line of reasoning is to link apparently aspatial policies to regional change differentials by (qualitatively) identifying local structural conditions that influence, or determine, policy impacts; then, to demonstrate statistically that the incidence of these conditions relates to observed differentials among cantones.

Variables representing structural conditions

Identified here are canton-level variables representing structural conditions that mediate, or determine, impacts of commodity pricing, agricultural credit, and exchange rate policies.[5] These

policies are elements of a broader import substitution industrialization program, even though they have been discussed largely in
terms of agriculture. Accordingly, structural biases uncovered
thus far are symptomatic of more pervasive biases in the Ecuadorian economy, biases that extend to other policies and to macroeconomic forces in general. It is thus argued that variables such as
those employed in statistical analyses below are pertinent to
overall economic change.

In mediating policy, a basic structural characteristic is the
rural or *urban character* of a canton (Figure 7.2, Box 1), indicated here by variables denoting percent of the labor force whose
occupation is farming, percent employed in the agricultural sector,
percent in mining, and percent employed in urban-based economic
sectors (manufacturing, commerce, service, finance) (Table 7.3).
Import substitution industrialization overall favors urban areas, as
do specific policies examined above. Commodity pricing, for
example, provides inexpensive food for urban workers and alters
relative price levels in favor of urban interests, shifting domestic
terms of trade accordingly. Agricultural credit policies also affect
terms of trade within Ecuador by creating a situation whereby
traditional agriculturalists producing for domestic markets pay
above average interest rates and are, therefore, less competitive.
Alternatively, credit policies favor farm and mineral exploitation
operations using more contemporary, capital-intensive production
systems; these entities also are more closely aligned with, and
contribute to, urban interests. Finally, exchange rate policies
benefit urban economies by reducing relative costs of raw materials, capital goods, luxury items, and food. In general, then,
cantones that are more urban in composition should have experienced positive changes from 1974 through 1982.

Another structural characteristic relevant to policy impacts is
educational attainment and ability to read and write, represented
here by percent of the labor force that is literate (Figure 7.2, Box
2; Table 7.3). Illiterate farmers unable to complete the necessary
forms, or apprehensive of doing so, are screened out by loan
application procedures associated with Ecuador's credit policy.
Illiterate farmers also have less (and less accurate) information on
interest rates, price supports, location of ENAC outlets, and the
like. Hence, fewer benefits from Ecuador's price and credit
policies accrue to cantones with low literacy rates. Similar
observations apply to non-agricultural sectors. In general, then,
cantones with lower levels of educational attainment should have

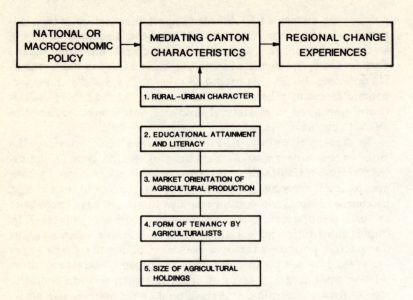

Figure 7.2 Structural conditions mediating policy impacts
among cantones

experienced less favorable changes for the period 1974–82.

Within agriculture, an important structural characteristic is the
market orientation of production (Figure 7.2, Box 3), indicated
here by percent of farm area devoted to annual crops (domestic),
percent devoted to perennials (export), and percent of farm area
under grass (livestock, especially cattle and dairying) (Table 7.3).
Price policies were more beneficial for export crops and dairying,
as were credit policies. In addition, export and cattle-dairy
production are often under capital-intensive systems employing
fertilizers, herbicides, improved seeds, specialized feeds, mechanization and the like; inputs which were cheapened by monetary
exchange policies overvaluing the sucre. In general, then, cantones oriented to export and livestock production, compared with
ones producing for domestic markets, should have enhanced their
position between 1974 and 1982.

The form of *agricultural tenancy* (Figure 7.2, Box 4) also
affects policy impacts, represented here by percent of farms
owned with title, held without title, rented with cash, and rented
by sharecropping (Table 7.3). Proof of land ownership is requi-

Table 7.3 Mean values for variables representing structural conditions of canton groups in 1974

Structural conditions and representative variables	Major urban cantones	Regional center cantones	Commercial agriculture cantones	Domestic agriculture cantones
Rural–urban characteristics				
Percent of the labor force				
whose occupation is farming	8.30	54.80	73.17	70.26
employed in agriculture	8.33	54.37	72.41	70.56
employed in mining	0.43	0.84	0.23	0.21
employed in urban-based economic sectors	67.54	33.11	17.16	21.76
Educational attainment and literacy				
Percent of the labor force literate	77.28	74.71	73.94	71.83
Market orientation of agricultural production				
Percent of farm area				
under annual crops	8.47	12.37	9.97	20.87
under perennial crops	5.06	8.03	21.17	5.98
under grass	28.15	37.01	32.80	41.46
Agricultural tenancy				
Percent of farms				
owned with title	53.06	50.82	52.00	56.61
held without title	4.86	17.12	12.22	10.48
rented with cash	3.66	2.54	4.94	2.27
rented by sharecropping	4.01	2.17	2.48	5.38
Size of agricultural holdings				
Percent small farms (<5 ha.)	66.37	53.43	48.39	64.71
Percent medium farms (5-50 ha.)	24.71	31.31	42.84	27.64
Percent large farms (51-500 ha.)	7.57	14.92	8.26	7.40
Percent very large farms (501+ ha.)	1.35	0.34	0.50	0.26
	n=2	n=36	n=22	n=49

site to obtaining BNF loans for agricultural production, thus excluding from credit policy benefits farmers who are without title or renting. Similarly, since sharecroppers produce minimal amounts for commercial markets, they would gain little from price support programs. Finally, because farmers renting, sharecropping, or without title tend to employ traditional agricultural methods, benefits from Ecuador's exchange rate policies would be limited. Hence, cantones characterized by farms owned with title, compared to those with other tenancy arrangements, should have changed in more beneficial ways during the period 1974–82.

The final set of structural variables indicates the distribution of *farms* among *size classes* (small, medium, large, very large) (Figure 7.2, Box 5; Table 7.3). A recurrent theme of regional change in Ecuador is that policy and macroeconomic impacts depend on an area's articulation with contemporary socioeconomic systems. National policies, conceived within such systems, assume (and/or strive for) an achievement-motivated, literate, and informed population; access to land, labor, capital, and technology; factor mobility; efficient marketing and communication channels; effectively operating supply, demand, and price mechanisms; and the like. In Ecuadorian agriculture, these characteristics are most prevalent where production tends towards capital intensity and use of wage labor, often but not exclusively for export markets. Such operations usually occur on medium and large farms. A different situation, less likely to benefit from national policies and macroeconomic growth, is found where farms are oriented to domestic market and/or subsistence production under more traditional, labor-intensive agricultural methods. These operations occur almost exclusively on small farms.

Statistical analyses

Structural variables identified above are now employed to statistically discriminate between canton groups that show distinct patterns of socioeconomic change from 1974 through 1982 (Appendix 7.2). To briefly review characteristics of each group, the Major Urban entities of Quito and Guayaquil exhibited high levels of growth in absolute employment and net out-migration by the economically active, particularly male segment of the labor force. Regional Center cantones, scattered throughout Ecuador, experienced favorable shifts in economic structure including, among Sierra cantones, reorientation of production systems to better conform with current socioeconomic conditions; high levels of in-migration also occurred in these more vigorous economies. Commercial Agriculture cantones, located entirely in the Costa, experienced crop diversification, but with little change in economic structure. Domestic Agriculture cantones, located largely in the Sierra, experienced out-migration by the labor force; this reflects conditions of overpopulation coupled with reorientation from a traditional socioeconomic system to wage labor (commodification) and urban-focussed activities. Overall, then, change occurred primarily in Regional Center and Domestic Agriculture cantones, but the former represents economic robustness, while the latter represents forced adjustment.

Table 7.3 indicates the average value of structural variables for each canton group, but Major Urban entities will be omitted from further consideration since Quito and Guayaquil's situation is distinct. In 1974, Regional Centers were noticeably more urban than Domestic and Commercial Agriculture cantones which, in turn, were relatively similar on this characteristic. All groups served both export and domestic markets, but Commercial and Domestic Agriculture cantones had a more pronounced orientation. Domestic Agriculture cantones, with former haciendas devoted to dairying, featured a slightly larger proportion of farm area in pasture. Renting with cash was somewhat more prevalent in Commercial Agriculture cantones; sharecropping in Domestic Agriculture cantones. Farms held without title were somewhat higher for Regional Center cantones, reflecting their location in the Oriente where spontaneous settlement had recently occurred, or in the Sierra where traditional agricultural systems had been most entrenched. Farms held with title show relatively little variation among canton groups, as does literacy. Finally, small farms were most prevalent among Domestic Agriculture cantones; medium-size farms among Commercial Agriculture cantones; and large farms among Regional Centers.

Table 7.4 presents results from stepwise discriminant analysis, using F=1.0 as a threshold for variable inclusion. Two functions were extracted. Function I distinguishes canton groups on the basis of a *Traditional–Contemporary Dichotomy*, generating nearly similar group centroid values for Regional Center and Commercial Agriculture cantones (-0.78, -0.73) but a widely divergent value for Traditional Agriculture cantones (+0.91). Having made that distinction, Function II then draws a *dichotomy* between *Stable*

Table 7.4 Stepwise discriminant analysis of regional center, commercial agriculture, and domestic agriculture canton groups on the basis of variables representing structural conditions in 1974

Independent variables	Zero-order F-statistics	Multiple variable F-statistics	Discriminant function coefficients I	II
Rural–urban characteristics				
Percent of the labor force				
whose occupation is farming	10.17	2.74	1.06	0.71
employed in agriculture	10.26	0.14		
employed in mining	2.93	0.13		
employed in urban-based economic sectors	10.35	2.81	0.67	1.16
Educational attainment and literacy				
Percent of the labor force literate	17.86	7.23	-0.72	0.35
Market orientation of agricultural production				
Percent of farm area				
under annual crops	7.87	1.46	0.35	-0.14
under perennial crops	13.60	2.50	-0.09	-0.49
under grass	2.12	0.43		
Agricultural tenancy				
Percent of farms				
owned with title	0.60	0.62		
held without title	1.26	4.35	0.63	0.23
rented with cash	2.52	1.13	-0.25	-0.12
rented by sharecropping	6.62	2.82	0.39	0.14
Size of agricultural holdings				
Percent small farms (<5 ha.)	3.52	0.10		
Percent medium farms (5-50 ha.)	7.17	1.33	-0.12	-0.48
Percent large farms (51-500 ha.)	2.83	0.11		
Percent very large farms (501+ ha.)	2.70	0.59		
	Wilk's Lambda		0.42	0.72
	Overall F statistic		5.56	
	Centroids of canton groups			
	Regional center cantones		-0.78	0.64
	Commercial agriculture cantones		-0.73	-1.10
	Domestic agriculture cantones		0.91	0.01

and *Dynamic Economies*, with a centroid of -1.10 for Contemporary Agriculture cantones, +0.64 for Regional Centers, and a neutral +0.01 for Traditional Agriculture cantones. The discriminant analysis may be interpreted as follows.

One type of change among Ecuadorian cantones pertains to modernization as a *catch-up* type of *economic adjustment*. Land reform was an important stimulus, affecting Traditional Agriculture cantones significantly more than others – hence, Function I's distinction between that group and the other two. To explore this further, consider the structural variables individually. Cantones distinguished by the traditional side of Function I were characterized (in 1974) by agriculture oriented towards domestic markets, farms held without title or rented through sharecropping, and employment in both urban and rural activities. These are largely Sierran cantones that were central to the hacienda economy, but forced to change by the combination of land reform, ISI/agricultural policies, and world economic conditions that worked against traditional activities. By contrast, cantones distinguished by the contemporary side of Function I were characterized by commercial agriculture, medium-size farms, often rented with cash, and high levels of literacy. These areas would be minimally affected by land reform, but among the primary benefactors of ISI/agricultural policies.

A second major change pertains to *raising* the *level* of *urban/industrial/modern sector articulation*. A major stimulus was import substitution industrialization and its attendant policies which, among more advanced cantones, affected Regional Centers more than ones characterized by Commercial Agriculture – hence, Function II's distinction between those two groups, given that Traditional Agriculture cantones were already accounted for. Another stimulus was petroleum, which especially affected Oriente cantones included in the Regional Center group. This had direct effects, such as growth in economic activity related to mineral exploitation and infrastructure creation, which in turn incited agricultural development and colonization by enhancing accessibility to a vast amount of vacant or under-used land.

To embellish the picture, again consider structural variables. Regional Center cantones, representing the more dynamic side of Function II, were characterized (in 1974) by economies with both rural and urban components, high levels of literacy, and farms rented by sharecropping or held without title. These cantones generally contain intermediate-size urban centers through which the economic transformation related to ISI/agricultural policies and petroleum exploitation was funnelled. Their rural hinterlands benefited accordingly, particularly in the Sierra where traditional agricultural elements were common. Commercial Agriculture cantones, representing the more stable side of Function II, were characterized by agriculture oriented to both domestic and export markets, although more the latter, and medium-size farms, often rented with cash. These are largely Costa cantones with commercial agriculture that benefited from ISI/agricultural policies, but not to the point of economic transformation.

Summary and concluding observations

Underlying Chapters Six and Seven is the premise that development in Third World settings is in large part an artifact of national policies or of occurrences (often cyclical) in the world economy. Further, impacts of state intervention and macroeconomic events, rather than being uniform across the space-economy, are conditioned by local socioeconomic structures. This is relevant because elements of regional structure are located so as to form areal complexes that are distinct in socioeconomic character. An apparently aspatial policy may, therefore, lead to marked differences in regional change, or development. This occurs both through direct effects on structural elements and indirectly through spatial differentiation in individual responses.

Both effects were demonstrated in Chapter Six. With regard to direct impacts, land reform in Ecuador primarily affected cantones with large landholdings, and more so in the Sierra than in the Costa. With regard to individual responses, place characteristics influenced whether circulation or migration was undertaken, and the preponderance of each.

This having been demonstrated, Chapter Seven focusses directly on regional change and its link with national policy. Patterns of change among Ecuadorian cantones between 1974 and 1982 are first established. This leads to a fourfold classification into Major Urban, Regional Center, Commercial Agriculture, and Domestic Agriculture cantones; each represents distinct regional change experiences. Attention then turns to policies that influenced change; namely, agricultural credit, agricultural pricing, monetary exchange rate, and more generally, import substitution industrialization measures. These are discussed in terms of differential effects on individuals, entrepreneurial activities, and other elements of regional structure. A basis is thus provided for identifying variables representing relevant aspects of a canton's makeup, which then are employed to discriminate among the four canton groups. Hence, a link between policy and regional change is established, both qualitatively and statistically.

In practice, however, appropriately accounting for uneven regional change needs to consider more than the juxtaposition of policy and place characteristics. Of particular relevance are elements emphasized by the mainstream of work concerned with policy and the spatial form of development; e.g., urban system characteristics, city size and accessibility, infrastructure proliferation, ratchet effects within the urban hierarchy, facilitators of diffusion, and the like. The geographic scale represented by 'mainstream' factors tends to be national or regional, whereas local place characteristics and heterogeneity within regions have been emphasized here. By viewing these approaches in terms of geographic scale, their complementarity becomes evident.

A related issue is the role of policy and other exogenous forces relative to those emphasized by development paradigms; or alternatively, the picture of Third World change presented in this book compared to paradigmatic portrayals. Relevant here is the present chapter's indication that appearances can readily mislead. Consider the finding that two types of change occurred in the Ecuadorian economy between 1974 and 1982 – a catch-up type of economic adjustment whereby cantones moved from traditional to contemporary economic systems, and an increase in the level of urban/industrial/modern sector articulation among the already more modernized cantones. Also pertinent is the spatial pattern of change which was spatially localized in that Major Urban areas were relatively stagnant, Regional Centers grew significantly in modern sector articulation, and Traditional Agricultural areas experienced catch-up. On the surface, these findings conform with core–periphery expectations, and one might readily conclude there had been a diffusion of development, polarization reversal, and autonomous growth as suggested by orthodox conceptualizations. But further analysis indicated otherwise. The engine of growth was not autonomously driven; more important was world demand for petroleum and national policies related to import substitution industrialization and land reform. Yet autonomous forces also were at work, both in themselves and in guiding policy; but to what degree?

Appendix 7.1 Component scores indicating socioeconomic change for Ecuador cantones, 1974–82

Province Canton	Component scores			
	I	II	III	IV
Sierra region				
CARCHI				
Tulcan	-0.01	-0.42	1.18	1.16
Espejo	-0.17	-0.29	0.81	-0.39
Montufar	-0.13	-0.87	0.81	-0.05
IMBABURA				
Ibarra	0.40	-0.34	0.67	0.42
Antonio Ante	0.18	-0.31	1.56	-0.55
Cotacachi	-0.33	-0.36	1.17	-0.53
Otavalo	-0.05	-0.08	1.20	-1.43
PICHINCHA				
Quito	6.12	0.51	1.06	-1.27
Cayambe	-0.12	0.07	0.30	-0.89
Mejia	-0.02	0.15	0.74	0.05
Pedro Moncayo	-0.34	-0.54	0.85	-0.99
Ruminahui	0.31	0.34	0.62	2.06
Santo Domingo	0.30	0.26	-1.15	0.33
COTOPAXI				
Latacunga	0.14	-0.44	0.66	0.35
Pangua	-0.33	0.39	-0.16	0.07
Pujili	-0.14	0.17	0.15	-0.54
Salcedo	-0.18	-0.09	0.68	-1.13
Saquisili	-0.07	-0.39	0.92	-1.13
TUNGURAHUA				
Ambato	1.00	0.15	1.04	0.99
Banos	-0.04	-0.21	0.74	1.56
Pelileo	-0.30	0.00	0.64	0.17
Pillaro	-0.25	-0.28	1.07	-0.86
Patate	-0.34	0.65	0.93	0.14
Quero	-0.34	0.97	-0.40	-0.67
BOLIVAR				
Guaranda	-0.10	-0.32	0.28	-0.20
Chillanes	-0.51	-0.39	0.30	-1.05
Chimbo	-0.30	-0.95	1.07	-0.67
San Miguel	-0.27	-0.57	1.12	-1.25

Appendix 7.1, continued

Province Canton	I	Component scores II	III	IV
CHIMBORAZO				
Riobamba	0.56	-0.20	0.67	1.79
Alausi	-0.13	-2.76	0.65	0.65
Colta	-0.24	-0.50	-0.38	-1.93
Chunchi	-0.40	-0.82	0.68	-0.74
Guamote	-0.26	-0.09	-0.70	-1.67
Guano	-0.26	-0.55	1.36	-1.16
CANAR				
Azogues	-0.39	-0.17	1.47	-1.10
Biblian	-0.43	-0.35	1.35	-1.61
Canar	na	na	na	na
AZUAY				
Cuenca	1.16	0.16	0.99	0.93
Giron	-0.32	-0.21	-0.12	-0.87
Gualaceo	-0.33	0.18	0.27	-0.97
Paute	-0.76	0.48	0.88	0.58
Santa Isabel	-0.54	-0.17	0.29	-0.17
Sigsig	-0.54	-0.45	0.18	-1.31
LOJA				
Loja	0.57	-0.72	0.10	2.46
Calvas	-0.56	-0.24	0.81	1.65
Celica	-0.65	-0.76	0.63	1.23
Espindola	-0.54	-0.09	-0.03	-0.86
Gonzanama	-0.57	0.52	0.22	0.55
Macara	-0.41	-1.48	-0.17	2.24
Paltas	-0.21	0.12	-0.06	-0.15
Puyango	-0.48	-0.89	0.25	-0.38
Saraguro	-0.48	0.08	0.61	-0.96
Costa region				
ESMERALDAS				
Esmeraldas	0.55	0.15	-0.74	0.59
Eloy Alfaro	-0.54	0.13	-0.48	-1.16
Muisne	-0.60	-0.41	-0.38	-1.17
Quininde	-0.34	-0.25	-0.32	-0.46

Appendix 7.1, continued

Province Canton	I	Component scores II	III	IV
MANABI				
Portoviejo	0.87	-0.27	-0.88	1.41
Bolivar	-0.16	-0.81	-1.63	-0.79
Chone	0.08	-0.29	-1.32	-0.32
El Carmen	na	na	na	na
Jipijapa	0.06	1.33	-3.59	1.25
Junin	-0.33	-0.63	-0.02	-0.64
Manta	0.70	-0.67	0.34	1.12
Montecristi	-0.06	-0.60	-1.30	-0.67
Pajan	-0.43	-0.31	-2.41	0.15
Rocafuerte	-0.12	-0.09	-1.33	-0.82
Santa Ana	-0.13	-1.01	-2.23	-0.74
Sucre	-0.07	-1.26	-1.76	0.25
24 de Mayo	-0.34	-1.22	-2.48	-0.06
LOS RIOS				
Babahoyo	0.41	-0.45	-1.07	0.68
Baba	-0.40	0.17	-0.66	-1.00
Puebloviejo	-0.22	0.61	-0.62	-1.05
Quevedo	0.33	-0.12	-0.84	0.06
Urdaneta	-0.27	-0.39	-0.58	0.01
Ventanas	-0.04	-0.06	-1.19	-0.27
Vinces	-0.17	-0.17	-0.02	-1.15
GUAYAS				
Guayaquil	7.39	0.01	-0.36	-1.29
Balzar	-0.09	-0.25	-1.29	-0.42
Daule	0.26	-0.15	-2.47	-0.13
El Empalme	-0.04	-0.51	-2.29	-0.35
Milagro	0.69	-0.29	-0.15	0.28
Naranjal	-0.16	0.08	-0.69	0.57
Naranjito	-0.16	0.34	-0.07	1.47
Salinas	0.26	0.66	1.29	0.98
Samborondon	0.08	0.41	-1.88	0.07
Santa Elena	0.17	-0.74	-0.49	0.05
Urbina Jado	-0.30	0.25	-0.12	-0.97
Yaguachi	-0.03	-0.45	-1.61	1.21
EL ORO				
Machala	0.92	-0.23	-0.24	2.12
Arenillas	-0.05	0.89	-0.57	1.62
Pasaje	0.04	0.10	0.72	2.18
Pinas	-0.20	0.70	0.10	0.62
Santa Rosa	0.13	0.31	0.61	2.23
Zaruma	-0.15	-0.88	1.00	0.43

Appendix 7.1, continued

Province Canton		Component scores		
	I	II	III	IV
Oriente region				
NAPO				
Tena	-0.31	0.31	-0.19	-0.99
Aguarico	-0.54	-0.46	0.98	0.44
Orellana	-0.42	5.10	0.50	-0.78
Putumayo	-0.26	6.27	-0.51	0.38
Quijos	-0.28	0.70	0.82	0.96
Sucumbios	-0.46	0.78	0.39	-0.36
PASTAZA				
Pastaza	-0.08	0.52	-0.11	0.14
Mera	-0.08	-0.24	0.40	1.50
MORONA SANTIAGO				
Morona	-0.36	2.29	0.34	-0.73
Gualaquiza	-0.29	1.19	-0.08	0.37
Limon Indanza	-0.39	-0.13	0.80	-0.05
Palora	-0.32	0.58	0.48	0.53
Santiago	-0.40	-0.03	0.74	0.65
Sucua	-0.41	0.60	0.11	-0.08
ZAMORA CHINCHIPE				
Zamora	-0.19	0.32	0.55	0.03
Chinchipe	-0.28	-0.08	0.27	0.50
Yacuambi	-0.45	0.74	-0.28	-0.32

Appendix 7.2 Ecuador cantones grouped according to socioeconomic change profiles for the period 1974–82, and related characteristics

Canton groups and members	Place characteristics in 1982			
	Popul'tn	Pct urban	Region	Comments

Major urban cantones

Quito	1,116,035	77.64	Sierra	National and provincial capital, principal commercial center
Guayaquil	1,328,005	90.31	Costa	Ecuador's principal port and commercial center, province capital

Regional center cantones

Tulcan	59,474	52.10	Sierra	Province capital, border town with Colombia
Ibarra	125,876	45.64	Sierra	Province capital
Ruminahui	32,537	46.37	Sierra	Proximate to Quito
Latacunga	125,381	22.94	Sierra	Province capital, commercial center for agricultural produce
Ambato	220,477	45.56	Sierra	Province capital, regional commercial center with artisannal production, particularly of shoes and foodstuffs.
Banos	14,575	57.22	Sierra	Tourism
Patate	9,605	16.73	Sierra	Agricultural production area
Riobamba	151,623	49.76	Sierra	Province capital, commercial center
Cuenca	275,070	55.41	Sierra	Third largest city in Ecuador, province capital, ceramics, artisannal production of jewelry, woven goods from vegetable fiber
Paute	36,178	6.46	Sierra	
Loja	121,317	59.06	Sierra	Province capital, regional commercial center
Calvas	34,886	29.06	Sierra	
Celica	26,258	16.27	Sierra	
Gonzanama	35,620	18.48	Sierra	
Macara	17,753	59.20	Sierra	
Portoviejo	167,085	61.42	Costa	Province capital, one of three major commercial centers in Costa
Manta	106,364	94.33	Costa	Second largest ocean port, linked to Portoviejo, much recent growth
Milagro	107,188	71.85	Costa	Commercial center, major agricultural production area
Naranjito	17,764	59.24	Costa	Proximate to Guayaquil
Salinas	67,941	26.12	Costa	Tourism, shrimp farming
Machala	126,492	86.49	Costa	Province capital, major commercial center, shrimp farming
Arenillas	42,502	68.97	Costa	
Pasaje	46,774	56.07	Costa	Major agricultural production area
Pinas	40,249	30.12	Costa	Major agricultural production area
Santa Rosa	42,262	63.22	Costa	Major agricultural production area

Appendix 7.2, continued

Canton groups and members	Place characteristics in 1982			
	Popul'tn	Pct urban	Region	Comments
Zaruma	36,594	27.83	Costa	Gold and other mining
Aguarico	3,241	8.79	Oriente	Petroleum extraction
Orellana	29,189	13.69	Oriente	Commercial center, petroleum extraction, channel for colonization
Putumayo	26,969	29.58	Oriente	Border area with Colombia
Quijos	9,175	3.80	Oriente	Crossroads area, agriculture
Pastaza	27,679	35.25	Oriente	Province capital, agriculture, channel for colonization, indigenous economic change
Mera	4,100	13.88	Oriente	Agricultural production area, channel for colonization, indigenous economic change
Gualaquiza	10,482	25.80	Oriente	Colonization, indigenous economic change
Palora	5,395	33.68	Oriente	Colonization, indigenous economic change
Santiago	7,612	16.70	Oriente	Colonization, indigenous economic change
Chinchipe	8,733	21.65	Oriente	Colonization

Commercial agriculture cantones

Santo Domingo	138,065	50.15	Sierra	Sierra province, but town is in the Costa, major commercial center with rapid growth in last two decades, transportation hub, major agricultural production area
Esmeraldas	140,513	64.31	Costa	Province capital, oil refining and distribution point, third largest ocean port, one of three major commercial centers in Costa
Bolivar	58,371	16.33	Costa	Major agricultural production area
Chone	138,862	24.37	Costa	Major agricultural production area
Jipijapa	72,940	37.22	Costa	Artisannal production of woven goods from vegetable fiber, agriculture
Montecristi	31,793	25.57	Costa	Artisannal production of woven goods from vegetable fiber, agriculture
Pajan	41,521	11.82	Costa	Major agricultural production area
Rocafuerte	51,003	10.77	Costa	Major agricultural production area
Santa Ana	58,917	10.22	Costa	Major agricultural production area
Sucre	87,568	14.11	Costa	
24 de Mayo	36,271	11.16	Costa	Major agricultural production area
Babahoyo	106,628	39.64	Costa	Province capital, major agricultural production area

Appendix 7.2, continued

Canton groups and members	Place characteristics in 1982			
	Popul'tn	Pct urban	Region	Comments
Quevedo	164,920	40.64	Costa	Commercial center, major agricultural production area
Urdaneta	21,186	15.83	Costa	Agricultural production area
Ventanas	50,779	31.25	Costa	Major agricultural production area
Balzar	58,616	30.07	Costa	Major agricultural production area
Daule	141,993	13.33	Costa	Major agricultural production area
El Empalme	52,619	32.34	Costa	Major agricultural production area
Naranjal	35,583	26.93	Costa	Major agricultural production area, shrimp farming
Samborondon	25,430	28.06	Costa	Major agricultural production area
Santa Elena	72,490	17.74	Costa	Tourism, fishing, petroleum extraction and refining
Yaguachi	90,192	7.62	Costa	Major agricultural production area

Domestic agriculture cantones

Espejo	26,030	22.98	Sierra	
Montufar	42,275	26.52	Sierra	
Antonio Ante	26,339	46.49	Sierra	
Cotacachi	31,912	16.23	Sierra	Artisannal production of leather and textiles
Otavalo	63,160	27.66	Sierra	Artisannal production of textiles, commercial center
Cayambe	41,740	34.36	Sierra	
Mejia	39,016	16.73	Sierra	
Pedro Moncayo	14,732	12.48	Sierra	
Pangua	18,581	6.75	Sierra	
Pujili	76,868	4.97	Sierra	
Salcedo	42,004	14.03	Sierra	
Saquisili	14,844	19.62	Sierra	
Pelileo	40,111	11.42	Sierra	
Pillaro	31,565	13.49	Sierra	
Quero	14,177	8.90	Sierra	
Guaranda	72,917	18.77	Sierra	Province capital
Chillanes	20,129	9.82	Sierra	
Chimbo	23,991	13.48	Sierra	
San Miguel	28,912	13.36	Sierra	
Alausi	60,101	13.21	Sierra	
Colta	55,428	3.91	Sierra	
Chunchi	14,646	21.81	Sierra	
Guamote	25,362	8.87	Sierra	
Guano	42,433	14.46	Sierra	
Azogues	68,273	21.31	Sierra	Province capital
Biblian	20,955	15.35	Sierra	
Giron	35,306	7.59	Sierra	
Gualaceo	40,460	16.23	Sierra	
Santa Isabel	30,939	7.21	Sierra	

Appendix 7.2, continued

Canton groups and members	Popul'tn	Pct urban	Region	Comments
	Place characteristics in 1982			
Sigsig	24,066	12.20	Sierra	
Espindola	18,176	7.78	Sierra	
Paltas	59,246	17.78	Sierra	
Puyango	21,859	15.85	Sierra	
Saraguro	25,653	8.13	Sierra	Artisannal production of textiles
Eloy Alfaro	46,003	30.18	Costa	
Muisne	16,746	21.86	Costa	Growing tourist area
Quininde	45,746	23.30	Costa	Agricultural production area
Junin	17,903	19.75	Costa	
Baba	27,299	51.24	Costa	
Puebloviejo	18,929	20.39	Costa	
Vinces	66,128	22.09	Costa	
Urbina Jado	40,633	12.13	Costa	
Tena	41,071	17.46	Oriente	Province capital, commercial center, channel for colonization
Sucumbios	5,465	4.26	Oriente	
Morona	23,730	21.13	Oriente	Province capital, commercial center, focus for colonization
Limon Indanza	10,734	21.53	Oriente	Colonization, indigenous economic change
Sucua	12,299	30.50	Oriente	Indigenous economic change
Zamora	21,602	24.39	Oriente	Province capital
Yacuambi	16,356	20.84	Oriente	

8 Third World development as the local articulation of world economic and political conditions, donor-nation actions, and government policies: Concluding observations

Research reported in this book was initially motivated by a concern with development effects on migration and cognate processes in Third World settings. Representing that focus are studies of aggregate migration flows in Costa Rica (Chapter Three), individual population movements in Venezuela (Chapter Four) and Ecuador (Chapter Six), and individual labor market experiences in Venezuela (Chapter Five). These convincingly establish that place characteristics associated with development play a major role in societal processes.

For example, Chapter Three's examination of aggregate migration flows demonstrates spatial variation in the effect of variables pertaining to employment opportunities and information levels, which is traced to development events affecting particular locales and not others. Such events included world price increases for sugar cane and cacao, introduction of disease resistant banana varieties, government policies to modernize and stimulate cattle production, and policies/donor-nation actions related to completing the Pan American Highway link with Panama.

Additional findings are provided by Chapters Four and Five which account for individual out-migration, educational attainment, labor force participation, and wages received in Venezuela. Among factors considered, the traditional–contemporary character of a person's place of residence was most important, and somewhat significant was a locale's population pressure and/or the degree to which its population is dependent, rather than economically active. These place characteristics also affected the role of personal attributes such as gender, age, and educational attainment. Overall, for example, females were more likely to outmigrate than males, but this relationship was considerably stronger in less developed settings.

Underlying the role of development in Venezuela were policies to spatially decentralize education and employment opportunities. That the former was more successful aggravated already existing disparities between the distribution of employment opportunities and the educated work force, which increased propensities for urban-directed migration, particularly among females.

Focus turned directly to policy in Chapters Six and Seven. These consider the role of land reform in population movements from Ecuador's rural Sierra and, for regional change over the whole of Ecuador, the role of several measures related to import substitution industrialization. Both chapters examine the manner by which national policies and other exogenous forces are articulated in local areas, reveal high variability in policy effects among places, and show that an important determinant of effect is local socioeconomic structures or place characteristics associated with development, which function as mediating agents.

The transition in focus just summarized grew out of employing development context as an instrumental variable, which forced attention to the question 'What is development in Third World settings?'. Because Third World development has been a topic of scholarly writing for more than three decades, including an extensive political economy critique, this aspect of the research was initially approached as a straightforward task. Indicative is the author's Development Paradigm of Migration, reported in Chapter Three, which adopts established views of development in an uncritical manner.

It became increasingly evident, however, that established (or conventional) conceptualizations of development, synopsized in Chapter Two, were not suitable for understanding the role of place in population processes at the local level, where investigation was focussed. Accordingly, the question 'what is development', originally viewed as a background exercise, became a major topic.

Formulating an appropriate framework proved difficult. A means of resolving contradictions between established views of development and the author's perception of Third World reality grew out of research on migration and related processes reported in earlier chapters. But the framework itself became evident only in the course of writing this book, when findings of the various research modules were set beside one another. Refinement is still ongoing.

The current version, presented in Chapter Two, advocates that development aspects of Third World landscapes and their change over time represent the local articulation of world economic and political conditions, donor-nation actions, and policies of Third World governments themselves. Spatial variation in the manifestation of these forces occurs through their interaction with local conditions. Hence, to understand development, or regional change, an appropriate analytical framework considers the aforementioned external, or exogenous, forces; the way they are mediated by local conditions; net effects on enterprises and individuals (and/or their response); and how the conjunction of these elements translates into socioeconomic change at local, regional, and national scales.

Considerations related to understanding development in Third World settings[1]

Issues stemming from, or related to, the book's perspective on Third World development comprise the final concern. In exploring these, studies of population movements, related societal processes, and regional change reported in earlier chapters are an essential ingredient, providing both insights and illustrations. These studies primarily focus on national policies, but their analytic approach, reasoning, and conclusions apply equally well to donor-nation actions and world economic and political conditions.

A central theme of the book is the need to move away from inquiry organized around established development frameworks. Proposed instead is a less nomothetic, more *idiographic* perspective that takes account of places as entities in their own right and their unique experience of change. Particularly important is an ongoing dialogue with *place reality*, and continual revision to align one's viewpoint accordingly. Elements of orthodox and political economy frameworks are neither precluded nor assumed. But their role needs to be tempered in terms of the situation being considered and exogenous forces that directly impact on and shape place dynamics.

This approach reflects discontent with customary research strategies. In the author's case, disquiet stems from frustration in trying to mesh his experiential sense of the Third World with its social science portrayal. But as noted in Chapter Two, others with dramatically different research orientations have voiced similar concerns and urge a shift away from strongly nomothetic objectives. The present chapter, and this book, represent a step towards solidifying that thrust.

Discussion proceeds in three directions. First, Statistical Representations of Development, Place Knowledge, and their use as explanatory elements are considered. Attention then turns to the broader topic of Regional Change and Research Protocols for Studying It. The chapter closes with a set of Summary Observations and Research Implications.

The statistical representation of development and place knowledge

This book's argument provides a rationale for *scaling* procedures originally employed to delineate modernization surfaces and their change over time. Associated with the modernization paradigm and its view of development as a diffusion process (Brown 1981: ch. 8), these procedures generate multivariate indices for sub-national areas on the basis of variables pertaining to educational opportunity, economic structure, urbanization, demographic profiles, infrastructures, government institutions, and the like. Such characteristics are manifestations of, or directly represent, world economic and political conditions, donor-nation actions, and policies of Third World governments. Further, deriving scale indices for sub-national units indicates variation in local articulation. Application of this approach is demonstrated in Chapter Four for Venezuela.

But scaling procedures are only a first step to understanding development impacts on local areas, not an end in themselves. In the Venezuelan application, for example, the scales STRUCTURE and PRESSURE are highly general and, if taken literally, say little about the mechanisms or processes being represented. Accordingly, as a precursor to depicting the operation of development in other societal processes, the meaning of scale indices in particular locales must be delineated. *Critical* to this task is *knowledge of place.*

An example is provided by Chapter Six, which examines individual out-migration and out-circulation from Ecuador's rural Sierra as a function of personal attributes and place characteristics related to agrarian structure. Indices summarizing place characteristics distinguish cantones on the basis of recency of landholdings, production for semi-subsistence or market purposes, size of landholdings, and the modernity of their socioeconomic environment. But understanding the role of these characteristics also required knowledge of Ecuador's ongoing transition from feudal to commodified labor and production structures; its land reform policies which were a component of the transition; how policy articulation varied in different rural settings; and details on cantones themselves at ground level.

Without such knowledge, that is, drawing on statistical findings alone, one might conclude that out-migrants originated from small farms and out-circulators from larger farms. In fact, both originated from small farms, but under different social and economic conditions at the local level. Circulation reflects identification with a cultural or ethnic community and holding land (usually a small parcel); it may be a response to recent circumstances or a long-standing community tradition; and it may involve artisannal activity, urban enterprise, or agricultural employment. Migration is associated with contemporary (rather than traditional) production systems; opportunities to settle new lands; and accepting the transition to labor as a commodity. Playing a role in both circulation and migration were limited land, land deterioration, and other aspects of population pressure; conflicts between social classes and/or communities; natural disasters such as the 1968 drought in southern portions of the Sierra; and land reform itself, which created economic pressures but also provided mechanisms by which land title could be obtained.

Chapter Six employs a methodology that accentuates the socioeconomic heterogeneity of Ecuador's rural Sierra, both to provide a foundation for generalization and to counter social science tendencies towards oversimplification. But it delineates landscape heterogeneity, using as a focal point the local articulation of forces highlighted in this book. Centering the analysis in this manner provides not only a basis, but also a *direction*, for generalization.

The issue of regional change and research protocols for studying it

Research on out-circulation and out-migration of individuals contributes to a broader concern with *regional change*, a fundamental topic of geographic inquiry. It already has been emphasized that delineating this process needs to center on the local articulation of exogenous forces. That these be considered in their *own right* also is supported by the argument. At issue is the choice (and emphasis) of elements comprising a study, which should be governed by substantive aspects of the research question rather than an established paradigm. Hence, to understand the ground-level configuration of movements from Ecuador's rural Sierra in a *tangible* and concrete manner, and avoid diversion by factors of minimal relevance in that context, Chapter Six disregards dependency aspects of Ecuador's patron/hacienda system, neoclassical economic aspects of utility-maximizing behavior by peasants, and the presence of both elements in economic motivations underlying land reform policies.

Treating exogenous catalysts in their own right has another consequence. Such forces play a fundamental role in shaping today's Third World, but we *know little* about their operation in a *spatial* frame of reference. Among many examples, petroleum as both an export and import has tremendously affected nations as a whole and regions within nations. A similar observation applies to other commodities, Third World debt crises, and import substitution (or export-led) industrialization policies which have been adopted in some form by the majority of Third World countries. By directly addressing such phenomena, our understanding of Third World change will be extended.[2]

Examples illustrating the approach advocated here include the following. Among *cross-national* comparisons, Roberts (1978) shows how policies, social alignments, macroeconomic events, and other forces related to economic expansion over several centuries (but largely the last) had different effects on Argentina, Brazil, Mexico, Peru, Britain, and the United States; Armstrong and McGee (1985) examine post-1940s economic expansion in a similar vein, comparing urban systems of Latin America and Asia; Amin (1977) considers colonial impacts on tropical Africa; Watts and Bassett (1986) contrast agrarian development in Nigeria and the Ivory Coast.

Less common are studies of regional change differentials *within* a nation, as in Helmsing (1986) on Colombia; Smith, Huh, and Demko (1983) on South Korea; or Lawson (1988) and Chapter Seven on Ecuador. Examined for Ecuador are the effects of land reform and elements of its import substitution industrialization program such as agricultural pricing, agricultural credit, and monetary exchange rate policies. Change is delineated at the canton level; then accounted for by the interaction of policies and place characteristics that mediate policy impacts. Land reform, for example, most visibly affected cantones with large landholdings, and among those, Sierra locales with more traditional agrarian structures. Agricultural pricing had deleterious effects on cantones oriented towards domestic market production; agricultural credit policies had greater benefits where agrarian structure was more contemporary. Policy impacts also were affected by a canton's proximity to urban centers or traditionalism of its local economy. Hence, apparently aspatial policies created regional change differentials, both directly through varying effects on canton structure and indirectly through spatially distinct effects on local enterprise and individual behavior.

Ecuador is not an anomaly. Government policies, macroeconomic forces, and donor-nation actions are referred to as *apparently aspatial* in the sense that they lack an areal focus but, nevertheless, have highly variable place-to-place effects. This occurs because their spatial articulation is *conditioned* by local socioeconomic structures, the elements of which (persons, households, entrepreneurial entities, urban agglomerations, infrastructures, resources, production possibilities, and the like) are located so as to form areal complexes that are distinct in character. By similar reasoning, *explicitly spatial* policies, macroeconomic forces, and donor-nation actions also exhibit spatial impacts differing from those anticipated.

Consider Venezuela's policies to spatially decentralize both educational opportunity and economic activity, which were initiated during the 1950s as part of a broad effort to disseminate petroleum wealth benefits. While educational opportunity was widely dispersed, raising educational attainment throughout the country, industrial decentralization was limited primarily to the Ciudad Guayana complex and urban areas near Caracas, the major city. This difference in the scale of decentralization exacerbated spatial disparities between employment opportunities and the educated work force, and separation from appropriate labor markets raised an individual's risk of being under- or unemployed. The net effect was to increase migration propensities already present in a highly polarized, urban-focussed economy, particularly among females. Chapter Five elaborates this situation.

The preceding discussion views place, or *endogenous*, characteristics as agents that condition (or channel) the local articulation of external, or exogenous, forces. By contrast, as noted in Chapter Two, orthodox conceptualizations depict endogenous characteristics in terms of autonomous development; i.e., self-sustaining, internally generated, independent growth. In the dual economy model, for example, development is driven by low-wage indigenous labor, which provides the basis for an ongoing expansion of industry; and in modernization theory, by self-perpetuating shifts in individual attitudes, leading to a more skilled, entrepreneurial, and achievement-oriented work force. The spatial manifestation of these autonomous development forces is a modern core (or growth pole) which gradually expands into, and engulfs, the traditional periphery (or hinterland).

Hence, this book's view of endogenous characteristics departs from that of orthodox thinking. It argues that the *imprint* of development on Third World landscapes largely reflects exogenous forces; that the impact of endogenous characteristics which might produce self-sustaining, internally generated, autonomous growth is, by comparison, negligible; but that endogenous factors are important in terms of their *interplay* with exogenous forces, leading to place variation in the aforementioned 'imprint'.

To elaborate, consider Chapter Seven's finding that two types of change occurred in the Ecuadorian economy between 1974 and 1982: a catch-up type economic adjustment whereby cantones moved from traditional to contemporary economic systems, and an increase in levels of urban/industrial/modern sector articulation among the already more modernized cantones. These changes were spatially localized in that major urban areas remained relatively stagnant, regional centers grew significantly in modern sector articulation, and catch-up occurred primarily in more traditional agricultural areas. Generally, then, a core–periphery landscape experienced the diffusion of development and polarization reversal. Because these *patterns* match ones associated with orthodox conceptualizations of development, autonomous endogenously driven growth might be assumed. But further analysis indicated the engine of change was world demand for petroleum, and national policies related to Ecuador's land reform and import substitution industrialization programs. At the same time, local (or endogenous) characteristics played a critical role by channeling (conditioning) the spatial articulation of external forces.

Consider another instance of apparently autonomous growth in economic activity. *Gold production* in Ecuador is soaring, accompanied by a shift from artisannal (e.g., panning by hand) to contemporary techniques of extraction.[3] Expansion has been engineered primarily by Ecuadorians, building on indigenous skills or, in more general terms, on an existing human resource base. This is exemplified by Zaruma, a small town in El Oro (!) province, where gold extraction has been an ongoing enterprise since the Spanish conquest of the late 1500s, including production through an international corporation that closed its facility in the early 1950s. These facts suggest endogenously driven, autonomous growth. But motivating the recent boom is a dramatic rise in gold prices on the world market (from $297 per ounce in 1982 to nearly $500 in 1988) and a decision by Ecuador's central bank to increase gold reserves. Hence, close examination reveals that exogenous forces of the type emphasized in this book were the impetus of economic growth, but its spatial articulation (i.e., greatly increased gold production in Zaruma and some other locales, less so elsewhere) reflects endogenous characteristics.

A third Ecuadorian example, reported in the June 20, 1988 issue of *Newsweek* (p. 39), is the production of Panama hats, for which world demand has increased tremendously in recent years. Panama hats constituted a major Ecuadorian export in the 1940s, but production levels slackened considerably between then and the recent boom, a shift attributable to external conditions. Locales benefiting from (or responding to) the currently high demand are found in southern Ecuador's coastal and Sierra regions, focussed on the urban areas of Guayaquil and Cuenca. These locales are long-standing centers of production that retain a skilled, albeit aging, labor force capable of producing quality Panama hats and an organizational structure for marketing them; a set of endogenous development conditions.

While these examples illustrate that regional change represents an interplay between exogenous forces and endogenous characteristics, they neglect that many aspects of the interplay are better described as a dialectic. And in this regard, endogenous characteristics *become forces*. Consider Robert's (1977) study of the Mantaro area of Peru's central Sierra, a progressive and economically vibrant region. The national government, as part of its overall development effort, established production and marketing cooperatives to pool resources, bring about economies of scale, obtain capital, provide technical assistance, increase personal incomes, and the like. But these cooperatives conflicted with long-established local institutions and economic entities that accomplished similar goals. Small-scale entrepreneurs, operating to maximize personal welfare, responded by subverting the cooperative effort to their own purposes. As a broader outcome of this conflict and its resolution, income-maximizing strategies tended to shift away from opportunities generated by the (previously dominant) agro-mining and private enterprise sectors; instead, strategies became oriented towards tapping resources of the state, particularly towards obtaining government employment and education. Turned aside, therefore, were both an ongoing (and indigenous) development impetus at the local level and a new impetus nationally, and the Mantaro area itself was weakened economically.

Another aspect of the interplay between external and endogenous forces is its complexity. Topics such as gold production or a selected set of government policies in Brazil (Browder 1987), Costa Rica, Ecuador, Peru, or Venezuela are sufficiently focussed that their intricacy yields to research. But when trying to understand the *full* range of forces underlying regional dynamics, the research task increases exponentially.

Illustrative of this is Helmsing's (1986) exhaustive examination of Colombia. From 1900 to 1950, the major impetus to economic growth was commercial coffee production, which in turn stimulated infrastructure development, industrialization, and centralization of government. From 1945 to 1967, import substitution industrialization policies and economic consolidation played a major, though not necessarily beneficial, role throughout the economy; while agriculture specifically was affected by new technology, urban and industrial demand for its products, food and production policies, producer associations, and rural upheaval known as the *Violencia*. Major factors affecting the period 1967–80 included import liberalization, internationalization of the economy, agro-industrial integration, and marginalization of the peasantry. This exceedingly complex panorama becomes even more so when considering the articulation of socioeconomic change in particular regions; for example, Helmsing's analysis of Antioquia (ch. 7). In the end, however, Helmsing is able to generalize Colombia's regional dynamics in terms of four forces: agro-industrial integration, consolidation and extension of indigenous economic enterprise, growth in foreign enterprise, and centralization of the state. Amalgamating these, he then identifies socioeconomic 'verticalization' as the dominant, overarching process of regional change; i.e., a shift from balkanized, semi-independent local economies whose principal interactions were with one another ('horizontal') to interdependent, hierarchically organized socioeconomic structures wherein regional autonomy was sharply reduced or eliminated.

Summary observations and research implications

The preceding discussion provides several observations pertinent to research protocols for studying regional change and/or development. These also serve as a summary of major points concerning the book's view of Third World development.

First, patterns associated with conceptual constructs are often observed; for example, core–periphery, urban primacy, class conflicts, and structural or institutional rigidities. While these conform with paradigmatic expectations, probing beyond the correlation reveals that underlying forces fit better with the perspective advocated here – that the imprint of development on Third World landscapes represents the local articulation of world economic and political conditions, donor-nation actions, and policies of Third World governments.

Second, this situation supports research protocols which move beyond paradigmatic thinking, disencumber our mind-sets so as to encourage cross-fertilization from other perspectives, and emphasize inductive, substantively informed inquiry. They also remind us that similar spatial patterns may result from different processes, sometimes referred to as equifinality (Chorley 1962). Because appearances are often misleading, it is essential that investigations test alternative hypotheses, question relationships rather than take them at face value, and both accept and look for complementarities among alternative explanations.

Third, these assertions are bolstered by earlier recognition of the need to modify research protocols; for example, Chorley's (1962) call for 'open system' thinking and Sayer's (1984) treatise on 'realism'. But the present essay differs in emphasizing, and ascribing a central role to, *place knowledge*. Such knowledge includes thoroughly understanding, in a substantive manner, both endogenous characteristics and exogenous forces. It also includes a sense of place that transcends statistical measures and relationships by recognizing their meaning varies in different contexts. An often used illustration is GNP per capita; more intricate examples include Chapter Six's use of scale indices pertaining to Ecuador cantones, and demonstration in Chapter Three that the applicability (and interpretation) of conventional migration models varies both within and among nations. More generally, place knowledge is seen as a *touchstone* against which empirical findings should ring true; whether the question is broad, such as socioeconomic change at regional or national scales, or more narrowly defined, such as out-migration/circulation from rural Ecuador.[4]

The preceding two paragraphs are complementary concerning research strategy. This is illustrated by Watts and Bassett's (1986) examination of food shortages in Nigeria and the Ivory Coast. Because these nations are outwardly similar in several respects (e.g., high rates of economic growth based on commodity exports and extensive economic intervention by the national government), analysis might cease with a simple correlation. But further inquiry shows that Nigeria and the Ivory Coast differ in internal political structure and in relationships between the government, socioeconomic interests, and the populace overall; accordingly, dynamics underlying their food shortages also differ.

Fourth, change is seen as the outcome of an interplay between forces external to a locale and characteristics of the locale itself. In particular, *apparently aspatial* exogenous forces lead to substantially different outcomes from place to place because their spatial manifestation is mediated by local socioeconomic structures, the elements of which form areal complexes that are distinct in character. By similar reasoning, *explicitly spatial* exogenous forces also exhibit unanticipated variations in spatial impact.

Alternatively, forces innate to a locale, including tendencies for internally generated autonomous change, are conditioned by their external environment. Appropriate, then, is an analytical framework that considers both exogenous and endogenous forces, the way they condition one another, the net effect on local enterprise and individuals (and/or their response); and how the conjunction of these elements translates into socioeconomic change at local, regional, and national scales.

Fifth, the preceding indicates that substantive generalizations are likely to be *conditional*, not absolute. It should be understood, however, that 'conditional' and 'idiosyncratic' are not synonymous. Consider Helmsing's findings for Colombia, summarized above. Four broad forces are identified, then amalgamated into an overarching process of socioeconomic verticalization. These forces have shaped the space-economy of many Third World nations, not just Colombia.[5] They also parallel the mercantile, industrial, and late (or economic consolidation) capitalism constructs by which Johnston (1982, 1984) characterized North American space-economies. In this sense, Helmsing's findings display generality.

But Helmsing's (and Johnston's) generalizations also are

circumscribed in that their applicability, or relevance, varies according to the historical and present-day circumstances of each place. Even a seemingly straightforward proposition is likely to be conditional in this manner. For example, the widespread finding that post-1950 land reform in Latin America led to rural emigration (Chapter Six; Peek and Standing 1982a) reflects its role in the transition from feudal to commodified labor and production systems; were circumstances otherwise, a different finding might prevail.

Sixth, the approach to Third World development advocated here concatenates with (but evolved independently from) a Developed World thrust known as locality studies or the new regional geography, which in turn is closely linked with realist approaches to social science (Jonas 1988; Massey 1983, 1984: ch. 5; Sayer 1984, 1985). One issue raised by this emerging focus is the danger of idiographic findings, given that generalizations are conditional.[6] Resolution requires a framework to which findings can be related and from which generalizations may emerge.

A step in that direction for Third World development is deciding which exogenous forces are most relevant and, for each, what geographic focus is appropriate to understanding their impact on areal units. For example, import substitution industrialization policies, export-led growth, International Monetary Fund loan practices, land reform policies, and the operation of major commodity markets such as petroleum have occupied center stage in shaping today's Third World. If these forces are one's primary concern, geographic focus will depend on which is being considered. To examine land reform effects on Ecuador's space-economy, Chapter Six selected cantones with a tradition of hacienda-based agriculture; whereas Chapter Seven's (and Lawson's 1988) study of agricultural policies focussed on a broader range of cantones representing variations in rural characteristics and articulation with the contemporary agrarian economy.

Alternatively, a geographic focus might motivate the inquiry. It then would be appropriate to select a cross-section of locales (e.g., on the basis of socioeconomic structure or economic production characteristics) and examine the differing effects on each of one or more exogenous forces.

Whether research focus is organized around exogenous forces, a cross-section of places, or some combination thereof, the task would be assisted by a scheme for *classifying* exogenous forces. Chenery (1984), for example, distinguishes policy instruments as

applying to production, consumption, labor, trade, natural resources, monetary, fiscal, or investment matters; and indicates the
price and quantity variables affected by each. From a regional
change perspective, policies that are explicitly spatial ought to be
distinguished from those that are not. The former include development strategies focussing on growth poles, colonization, primate
or intermediate-size cities, rural areas, specific regions, infrastructure expansion, and decentralization of educational or employment
opportunities. Examples of aspatial policies include those discussed by Chenery and, from this book, land reform, agricultural
pricing, agricultural credit, monetary exchange, and import
substitution industrialization measures.[7]

However exogenous forces are addressed, we will learn more
about their operation in a spatial frame of reference, their varying
impact on local areas, and the way they shape Third World
change, or development, at sub-national levels. This alone
constitutes an important increment to knowledge.

Seventh, seeing national policy, world events, or donor-nation
actions as critical elements of Third World development is not
new in itself, but because many view these as secondary to other
dynamics, the *emphasis is significant*. Stressing the *local articulation* of exogenous forces represents an even greater departure
from the usual approach. For example, the urban bias of policy
has been given considerable attention, but primarily as a broad
force rather than how it impacts particular locales (Lentnek 1980;
Lipton 1976; Todaro and Stilkind 1981). Similarly, research
associated with neoclassical and political economy conceptualizations tends to focus on paradigm-related issues of a general nature
and, in spatial matters, broad regional effects rather than local
ones.

By contrast, this book contends that place and its experience
of change (or stasis) should be a central element of research
protocols; and that exogenous forces should be considered in their
own right, rather than in terms of neoclassical or political economy thinking. Generalization remains the objective, but it ought
to emanate from, not be imposed on, the locale or the substantive
process being studied. In this regard, we might heed the observation of Hirschman (1984: 88), a pervasive presence in development studies for thirty-five years.

I went to Colombia early in 1952 without any prior knowledge
of...economic development. This turned out to be a real

advantage; I looked at reality without theoretical preconceptions of any kind... [Later, I] discovered I had acquired a point of view of my own that was considerably at odds with current doctrines.

Eighth, in arguing that research on Third World settings should be grounded in and guided by the specifics of place, *place knowledge* becomes an essential ingredient. This emphasis has an association with traditional regional geography and its descriptive orientation towards informing on areal characteristics for their own sake. But quantification, spatial modeling, neoclassical reasoning, and political economy perspectives of the 1960s, 70s, and 80s (which may be seen as a reaction to traditional geography) also are represented. Advocated here then, and earlier by Taaffe (1974, 1985), is a return to the geography of place; not as an end in itself, but as a means for understanding societal processes, human behavior, and the role of place therein.[8]

To indicate the benefit of this approach, consider Third World labor markets. Research from a paradigmatic perspective might demonstrate how these exemplify the spatial or international division of labor whereby certain types of employment are allocated to Third World (and other) locations to take advantage of labor costs (e.g., Sayer 1986); alternatively, that construct might be used to interpret a specific labor market(s). But employment related to this phenomenon is only an element of the total picture, and in many Third World locales, a small one. Hence, rather than using a framework that is partially applicable at best, focussing on place as a totality should yield a comprehensive understanding that includes the role of spatial and international divisions of labor in Third World settings, local aspects of labor market dynamics *obscured* by a pre-defined research perspective, and how both vary from locale to locale. In this context, then, Wilber and Jameson's (1988: 25) exhortation towards

throwing off the conceptual blinders of the paradigms

means being less concerned with the spatial-international division of labor *per se* and more concerned with place. The appropriate question is not 'How does place i exemplify the spatial division of labor?' so much as 'What are the labor market dynamics of place i?'.

This book began with a question of the former type by focussing on the role of development in migration processes wherein Costa Rica was merely an illustration (Chapter Three). By Chapters Six and Seven population movements and development, or regional change, were still the concern, but focus had shifted to the place in which they occurred, Ecuador. The methodology also shifted. At both ends of the spectrum empirical analyses were based on census data providing broad geographic coverage and qualitative information, from either secondary sources or personal experiences, to interpret statistical findings. But as place became more important, so did the role of qualitative knowledge.

The next step might focus on small areas and specify the interaction between exogenous forces and local conditions in greater detail. Doing so would lead to a better understanding of, for example, why places respond differently to a similar exogenous force, or similarly to different forces.

This genre of inquiry usually has been directed towards providing a basis for improving policy or programmatic initiatives. Illustrations include Rudel's (1983) examination of colonization in the Ecuadorian Amazon and Reinhardt (1987) on efforts to modernize peasant agriculture in a Colombian village.

An alternative is using case studies to *inductively build generalizations*, or a broad understanding of conditions related to different/similar local responses. Bilsborrow's (1987) study of population pressure effects on rural areas provides an example, as does Helmsing (1986) on Colombia, and their approach is more congruent with themes elaborated here.

This complements the earlier observation that, too often, research tends to focus either on general constructs, wherein place (or a particular exogenous force) becomes a pawn in the argument; or on place/exogenous force(s) to the exclusion of deriving generalities. A conjunction of these foci is central to the research strategy advocated by this chapter, and the book. Highlighting similarities between places has been important, and conventional development paradigms are a segment of that task. But a more current need is to focus on place differences in order to gain a better understanding of local variation and its role in Third World development (or change) at all spatial scales. Ultimately, generalizations should emerge, and that remains as an objective – but generalizations which are rich in detail, recognize the *heterogeneity* of development processes, and emphasize *visible* mechanisms underlying regional change.

Said another way, considering Third World locales as totalities, and reviving thereby a long-standing geographic tradition, should increase our knowledge and lead to generalizations that build on, augment, and integrate those already available. The mechanism for accomplishing this is a *research strategy* that focusses on the *intersection* of external forces and local characteristics, draws on detailed knowledge of both, and seeks to provide generalizations in the presence of specific conditions. Such a tactic implements a *fusion* of approaches, blending the nomothetic perspective of development paradigms and the idiographic perspective identified with traditional regional geography.

Notes

Chapter 1: Introduction

1. Political economy perspectives include dependency, Marxist, and related historical-structural approaches. These terms are taken as equivalent, and used interchangeably, throughout the book.

Chapter 2: What is Third World development?

1. This section is a restatement of portions of research originally reported in Brown (1988). Permission for use obtained from *Economic Geography*.

2. As an analogy, consider economic-man precepts, a broad-gauged framework, and individual behavior. If one's primary interest is the economic-man model as an explanation of human behavior, that can be addressed by demonstrating congruity between the model and personal actions or characteristics. But if understanding a particular individual or group of individuals is the goal, economic man provides only a portion of the explanation and a more multifaceted, person-based perspective is warranted.

3. This classification captures the essence of paradigmatic thinking as it relates to the chapter's argument and, thereby, serves a heuristic purpose. Masked, however, is a highly heterogeneous set of viewpoints and philosophical positions.

4. Related to this observation is the role of interest groups in shaping the trajectory of development. In Argentina, for example, urban industrial interests and organized labor wrested political dominance from large rural landholders in the 1930s. But to form and maintain this coalition, wage levels and government programs affecting workers were continually upgraded, thus stoking inflationary pressures (Roberts 1978: ch. 3). Similarly, as shown in Chapter Seven of this book, Ecuador's economic policies of the 1970s bolstered urban industry and petroleum operations; deleteriously affected agriculture; but within agriculture, disfavored traditional more than contemporary operations.

5. Underlying observations of the preceding four paragraphs is the distinction between economic development and economic growth. Because these aspects of change are sometimes linked to one another, but sometimes not, the common tendency to use the terms synonymously leads to confusion. For further discussion, see Caporaso and Zare (1981), Flammang (1979), and Stohr (1982).

6. Evidence of this in a United States context includes Weber's (1899)

treatise on the growth of cities in the nineteenth century, emergence of the Chicago School of urban sociology in the 1920s (Short 1971), and recognition in the late 1970s of the Sunbelt's burgeoning economy (Berry and Dahmann 1977; Biggar 1979).

7. Earlier discussion around Figures 2.4 and 2.5 provide an example of such complementarity, as does Simon (1984) on understanding Third World colonial cities. Another example is rationales to account for the spatial-international division of labor. Political economy arguments build on basic relationships between capital and labor; orthodox-neoclassical arguments focus on innovation, technological change, product-life cycles, and cost-minimization strategies related to economic competition; but recent rationales blend these two approaches. For a fuller discussion, see Caporaso (1987: esp. chs. 1, 7), Frobel, Heinrichs, and Kreye (1980), and Markusen (1985, 1987). Also interesting is Scott and Storper (1986) which indicates the considerable degree that economic geography in a Developed World context has moved beyond debates between neoclassical/orthodox and political economy perspectives, and incorporated elements of both.

8. Armstrong and McGee is not an isolated example. Others include Dore's (1988) study of the Peruvian mining industry during the present century; Roseberry's (1983) examination of socioeconomic change from colonial times to the present in a coffee-producing locale of Venezuela, and Weeks (1985) study of industrialization in Peru from 1950 through 1980. With regard to the transition of geographic thinking itself, see Johnston (1987) for a general account and Forbes (1984: chs. 1–6) for Third World development studies.

9. This statement represents the generally accepted view that laissez-faire principles guided government actions until the presidency of F.D. Roosevelt. But recent historical research argues that government powers were used for economic development purposes since the nation's beginning; see Bourgin (1989) and Schlesinger (1986: ch. 9). This new interpretation of history may be relevant when themes outlined in the present chapter are further articulated.

10. This is illustrated by comparing Sinclair Lewis' (1922) *Babbit* with Steinbeck's (1939) Joad family in *The Grapes of Wrath*. A more specific and excellent account of the differential effects of socioeconomic structure on individuals is *Newsweek*'s Fiftieth Anniversary Issue (*Newsweek* Staff 1983). This traces Springfield, Ohio (chosen for its typicality) from the late 1800s to 1983, largely in terms of five families. The Bayleys are wealthy local industrialists whose firm was absorbed into a larger, nationally-based corporation in the 1960s. The Grams, originally farmers outside Springfield, shifted to a local ice cream and dairy business prior to World War II; this was successful for some time but forced to liquidate in the 1970s. The Bacons are Blacks with southern roots who helped found Springfield's civil rights

movement. Bayleys, Grams, and Bacons resided in or near Springfield prior to 1900. A more recent arrival, in the 1910s, were the Nusses, who worked in blue-collar occupations. The woman of this couple had recently immigrated from Germany; the man was a first generation American who maintained strong German roots. The Cappellis, both immigrants, also moved to Springfield in the 1910s; they became wealthy through a flower business and, later, real estate; a classic immigrant success story. To place these biographical sketches, and Springfield's social history, in a broader, economic context, see Johnston (1982, 1984). His novel approach to urban geography discusses structural forces shaping United States development from the early 1700s to the present in terms of three eras – Mercantile Capitalism, which (depending on location) was a force through the late 1800s; Industrial Capitalism, a force from the 1820s to the Depression; and Late Capitalism, a force from the 1910s to the present.

Chapter 3: Aggregate migration flows and development, with a Costa Rican example

1. To illustrate the contribution of Todaro's model to understanding Third World migrations, consider the rural-to-urban situation. Third World rural areas are characterized by high levels of under- or unemployment. Accordingly, the probability of securing a job in the rural area, $P_i(t)$, will be equal to (or near) 1.0 for the person already employed there, but very small for the unemployed. A person migrating to the city would face similar or only slightly better conditions for securing employment. However, putting aside that the wider variety of opportunities might itself be an inducement, urban wage levels, $Y_j(t)$, should be significantly greater than $Y_i(t)$ since rural areas are characterized by a near zero marginal productivity of labor; and the gap would be particularly marked for certain classes of persons such as the under- or unemployed. Given these conditions, the $P_j(t)Y_j(t)$ combination, even though small, would be greater than that same combination for rural areas. Hence, a rural resident with perfect knowledge of the urban labor market might rationally choose to migrate, and in the aggregate, a likely outcome is high levels of both urban unemployment and rural-to-urban migration (Todaro 1971, 1976: 28–36). This insight would not be possible through the labor force adjustment model.

2. The importance of migration chains can also be understood in terms of the active and passive migrant distinction of Hagerstrand (1957). Actives seek systematically for a new locale promising future prosperity. Passive migrants, who are considerably more numerous, follow and are dependent on impulses or information emanating from their active counterparts.

3. Migration chains are most commonly represented by place-of-birth

data wherein the number (or percent) of persons living in j but *born in i* is the independent variable, and the number (or percent) of persons in j who *previously resided* in i is the dependent variable. But time periods covered by these measures often overlap, so some movements are counted in both. To the extent this occurs, statistical independence is reduced and intercorrelation (falsely) increased (Kau and Sirmans 1979). One means of eliminating this is to use a migration chain surrogate, based on flows from an earlier, non-overlapping time period, as in Brown and Lawson (1985a, 1985b).

4. This scenario provides a rationale for policy initiatives to hasten or stimulate polarization reversal. For discussion see Renaud (1981), Richardson (1981), and Gore (1984).

5. Origin push is most commonly attributed to population pressure. Debated, however, is whether this results largely from conditions of society or conditions of nature. Illustrative of the former, a political economy contention, Gregory and Piche (1981) see population growth as a response to colonial systems and their contemporary remnants, while Fair (1982) attributes increased population pressure to disruption of traditional trading practices. Push from rural areas also has been explained by urban biases in development policies, which affect the locus of employment opportunities (Lipton 1976, 1980; Taylor 1980; Todaro and Stilkind 1981).

6. Note, however, Gardner (1981) who links contextual, or place, characteristics and individual decision-making, but does not apply his framework to Third World settings.

7. Another application of Zelinsky's framework is Fuchs and Demko's (1978) study of eastern Europe. Also interesting is Skeldon (1985), which considers the shifting role of circulation (and its distinction from migration) in relation to the mobility transition.

8. This is a restatement of research originally reported in Brown and Jones (1985). Permission for use obtained from *Demography*.

9. Other studies have estimated separate models for different internal migration streams; Greenwood, Ladman, and Siegel (1981), for example. But these give little, if any, attention to development–migration linkages.

10. This approach illuminates only one side of the symbiotic migration–development relationship; that is, development influences on migration are modeled, but not migration influences on development. Broadening the model could be accomplished by employing a simultaneous equations format with spatially varying parameters, an interesting step for future research.

11. In the application reported here, X–Y values are Cartesian coordinates representing centroids of Costa Rican cantones. These were derived by overlaying a grid on a Costa Rican base map, with 0,0 in the lower left-hand corner. Following Mather (1975), X and Y values were then taken as deviations from their respective means, a transformation which reduces intercorrelations between powers of X and Y variables in second-order polynomial models. The multicollinearity problem for higher-order surfaces also has been addressed by Jones (1984b) and Casetti and Jones (1987) using a technique incorporating principal components analysis.

12. Specifically, the probability of out-migration reflects their phase 'the decision to seek a new residence', and the probability of choosing j reflects their phase 'the decision of where to relocate'. For an alternative but parallel derivation of the dependent variable, and supporting rationale, see Rempel (1980, 1981).

13. Why origin push factors are masked by migration rate models is not entirely clear, but the following seems plausible. A large proportion of migration flows are targeted on the primate city (in the case of Costa Rica, San Jose), whereas no place has a comparable concentration of origins. Further, the primate city tends to have extreme values on independent variables; a much greater population, for example. Accordingly, when origin and destination factors are estimated simultaneously, as in migration rate models, j-relationships would be 'pulled out' by the primate city outlier, and i-relationships, with no comparable observation(s), would be overwhelmed. By contrast, using separate models for out-migration and relocation negates the statistical competition between these sets of variables.

14. Costa Rica's major political subdivision is the province, of which there are seven; its second-order subdivision is the canton. These numbered 79 in 1973, but because of census errors in tabulating migration flows, cantones were reconstituted to the 69 existing for the 1963 census. Further, cantones comprising the San Jose metropolitan area, the nation's capital and primate city, were treated as a single unit, leaving 59 observations. For details, see Brown and Lawson (1985b: Table 1). The 'five years of age or older' criterion is because the 1973 migration question was only asked of such persons.

15. Routine practice actually pertains to models estimated on the migration rate dependent variable (term (3.4)), but this is the joint probability of leaving i and relocating to j, as noted. Accordingly, logging variables of models pertaining to each component (terms (3.5) and (3.6)), and combining these in joint probability fashion, yields the logged variation of the migration rate model.

16. The most detailed articulation of the urban bias concept is Lipton (1976). A primary component of his argument is that Third World

research has dwelt on economic efficiency rather than equity, resulting in development strategies that point, for example, towards establishing agglomeration economies rather than satisfying basic needs. Such strategy translates into policy favoring urban over rural areas (Todaro and Stilkind 1981); hence the term urban bias. In migration research, urban bias is evident in the assumption that economic rationality underlies migrant behavior, and that aggregate migration patterns reflect the operation of equilibrating mechanisms. This emphasis has led to research designs which neglect additional aspects of migration, and generate findings more relevant to core, or urban, regions.

17. A more comprehensive cross-national comparison of this sort is provided in Jones and Brown (1985).

18. This model was obtained by allowing an additional step beyond the results reported in Table 3.2.

19. Ecuador's post-World War II land reform policies provide another example in which deleterious consequences accrued to the peasant (Brea 1986; Commander and Peek 1986; Peek 1982; Preston, Taveras, and Preston 1979). This is given further attention in Chapter Six.

20. Multicollinearity effects render distortion to the original map of SVPs associated with $PRES_j$. Hence, the second procedure was employed to get a clearer picture of population pressure effects across the landscape.

21. This finding also provides support for assumptions underlying Brown and Stetzer's (1984) simulation, which demonstrates that the migration–development link is related to, and has an effect on, urban system evolution. For a parallel discussion, see Ledent (1982).

22. Rural areas also supply the most migrants (60.1 percent) and San Jose the least (9.7 percent). The net result is a population transfer wherein rural areas emit 111,861 people, and receive 76,299; whereas San Jose emits 18,012 and receives 50,888. In numbers, then, rural areas are decreasing and San Jose is increasing (province capitals increase slightly), thus creating the impression of a rural-to-urban preference among migrants.

23. With regard to policy effects in other settings, Findley (1977, 1981), Pryor (1981), and Rhoda (1979, 1983) find that land reform, frontier-oriented resettlement schemes, and fertility control have reduced population pressure and, hence, rural out-migration. The opposite has occurred through diffusion of green revolution technologies, agricultural mechanization, and agricultural credit and extension programs, which generally favor better-endowed social classes, thus increasing social and economic disparities. Effects of irrigation programs have been mixed, depending on whether benefits are distributed in a

discriminatory or egalitarian manner. Rural-nonfarm activities have tended to slow rural out-migration initially, but gains in experience and skills often lead to an urban move later (Schneider-Sliwa and Brown 1986). Education in rural areas also has induced rural-to-urban migration in that it gives youth modern skills, attitudes, values, and knowledge. Social services have had no clear effect on rural out-migration. Rural out-migration also has been related to the level of social and economic disparity within a rural community, or changes therein (Gotsch 1972; Havens and Flinn 1975; Lipton 1980; Saint and Goldsmith 1980).

24. One might note the parallel between this framework and shift-share analysis which considers national, structural, and local effects on regional change. For a discussion, see Berry and Horton (1970: 98) and Paris (1970).

Chapter 4: Individual migration and place characteristics related to development in Venezuela

1. This is a restatement of research originally reported in Brown and Goetz (1987). Permission for use obtained from *Demography*.

2. As noted in Chapter Three, a framework addressing development influences on migration patterns was set out by Zelinsky in 1971, his hypothesis of the mobility transition; and in 1970 Mabogunje set forth a development-based conceptualization of rural-urban migration processes. But the topic did not attract extensive interest until the early 1980s; for example, Commander and Peek (1986), Findley (1981, 1982), Forbes (1981), Gardner (1981), Goldscheider (1987), Goldstein (1981), Hugo (1985), Morrison (1983), Peek and Standing (1979, 1982b), Rhoda (1979), Roberts (1982, 1985), Sabot (1979), Urzua (1981), and several studies by Brown and others beginning with Brown and Sander's (1981) 'development paradigm of migration'.

3. Ecuador is commonly divided into three regions: the Costa, comprising lowland areas bordering the Pacific; the Sierra, a highland area containing traditional population centers; and the Oriente or Amazon Basin. These regions are made up of provinces, the largest political unit, which in turn are subdivided into cantones. Quito, in the Sierra, is Ecuador's national capital.

4. Examples of this genre include studies of Chile (Berry 1969, Pedersen 1975), Ghana (McNulty 1969, 1976), Kenya (Soja 1968), Sierra Leone (Riddell 1970, 1976), and Tanzania (Gould 1976). At a broader scale of aggregation, Pedersen (1975) applies similar procedures to major political units (e.g., provinces, cantones) of South American nations,

treated collectively; Sheck, Brown, and Horton (1971) to major political units of the Central American Common Market nations, also treated collectively; and variations among nations of the world are examined by Adelman and Morris (1971), Berry (1961), and Tata and Schultz (1988). The preceding use factor analytic techniques. Another scaling procedure is exemplified by the Physical Quality of Life Index (PQLI), which measures basic needs fulfillment as an equally weighted average of infant mortality, literacy, and life expectancy at age one (Adelman and Morris 1973; Brodsky and Rodrik 1981; Hicks and Streeten 1979; Larson and Wilford 1979; Morris 1979).

5. An additional feature of multivariate scales is their use of readily available data and operational tractability. For an indication of the variety of variables that might be considered, see Ginsburg's (1961) Atlas of Economic Development or Todaro (1985: 21–59).

6. Seventeen variables were winnowed from a larger set in order to reduce redundancy and clarify component loading patterns. Concerning the unit of analysis, in 1971 Venezuela had 178 distritos, a local and usually small political division, which aggregate into 24 states or equivalent entities.

7. The CELADE sample comprises approximately five percent of the nation's 1971 population. This is one of many samples of individual records from national censuses of Latin American nations, made available by Centro Latinoamericano de Demografia (CELADE). For further discussion of CELADE data and a description of its 1970- and 1980-round data sets, see Centro Latinoamericano de Demografia (1974, 1978, 1984).

8. As can be seen from Table 4.1, more traditional settings, those with negative principal component scores, are indicated by a high percent of the population residing in rural areas, born in the same distrito, without schooling, and employed in primary activity. More contemporary structures, those with positive principal component scores, are indicated by a high percent of the population who are migrants, more educated, and employed in secondary and tertiary activity; also by higher averages in occupational status and wage levels.

9. As can be seen from Table 4.1, greater dependency, indicated by positive principal component scores, is related to larger families and to higher dependency ratios and population pressure indices. Lesser dependency, indicated by negative principal component scores, is related to the absence of these characteristics.

10. Since more favorable development conditions are indicated by positive scores on STRUCTURE (CS_1) and negative scores on PRESSURE (CS_2), they were combined as $CS_1 + (-1*CS_2)$. An alternative

formulation is to weight each component by its contribution to overall variance (for CS_1, 54.2%/66.1% and for CS_2, 11.9%/66.1%). Equal weighting was chosen for three reasons. First, this accentuates development aspects not associated with urban areas, which are often obscured by the strong urban bias of Third World space-economies. Second, the relative role of each component is addressed directly in statistical analyses of individual migrations and labor market experiences. Third, weighting is seen as a discretionary decision because the composite development index is primarily used in a descriptive, rather than statistical, manner.

11. Derivation of distrito variables was discussed in the previous section. It would have been preferable if these variables represented a time period preceding that of migration, but such data were not available. Hence, the model implicitly assumes that place characteristics motivating the observed migrations are distributed in a similar fashion to those occurring post facto, which is not unreasonable in an economy such as Venezuela's. This assumption also voids the issue of simultaneity bias (Greenwood 1975b); i.e., estimating contextual (or place) effects on migration without taking account of the simultaneous effects of migration on context. Further, simultaneity bias is more acute when both variables are aggregates and/or when context is a single variable directly affected by migration, rather than a composite variable. Hence, simultaneity is also offset by the research design employed here.

12. For a discussion of logistic regression see Aldrich and Nelson (1984) or Wrigley (1976).

13. Appendix 4.1 indicates which distritos had an urban center of 20,000 or greater population. Other distritos comprise the sample where no urban center equals or exceeds 20,000 population.

14. It should be kept in mind, however, that b_{dir} and b_{int} represent only a portion of the combined effects of STRUCTURE and AGE on outmigration; for the total effect, b_{dir} for STRUCTURE also must be considered. Further, b_{int} indicates both how AGE effects are modified by STRUCTURE, and how STRUCTURE effects are modified by AGE.

Chapter 5: Individual labor market experiences and place characteristics related to development in Venezuela

1. Data used here include the personal attributes age, gender, educational attainment, labor force participation, and wages received, any of which might be used to represent social category effects. In a developing

economy, however, persons of different age (even if living in the same place) should have experienced significantly different economic and social environments; hence, age reflects both development and social category effects. Educational attainment, labor force participation, and wages are attributes to be explained. In the context of this study, then, gender is the most appropriate social category.

2. As noted elsewhere in this book, that policy has an urban bias, often unintended, is a common occurrence.

3. This is a restatement of research originally reported in Brown, England, and Goetz (1989). Permission for use obtained from *International Regional Science Review*.

4. See, for example, Arizpe (1977); Beneria (1982); Beneria and Sen (1981); Boserup (1970); Boulding (1980); Buvinic, Lycette, and McGreevey (1983); Center for International Research (1985); Centro Economico Para America Latina (1983); Chinchilla (1977); Draper (1985); Fernandez-Kelly (1981); Jelin (1977, 1982); Mohan (1986); Recchini de Lattes (1983); Recchini de Lattes and Wainerman (1986); Standing (1982); Taglioretti (1983); Tiano (1986); Women and Geography Study Group of the IBG (1984); Zelinsky, Monk, and Hanson (1982).

5. Educational attainment ranges from 0 through 16 years. Wages received are in seven categories: less than 250 bolivares per month, 250–499, 500–749, 750–999, 1,000–1,499, 1,500–1,999, and 2,000 or more. Each individual in the sample was assigned the mid-point value of their respective wage category, except that '2,000 or more' received a value of 2,500. A 1971 bolivar was the equivalent of approximately 0.25 US$.

6. As noted in the preceding chapter, the CELADE sample consists of 439,815 records, representing approximately five percent of Venezuela's 1971 population. Regarding its use, questions motivating this study require data with broad geographical coverage (e.g., national) and individual records that include income as a variable. At the study's initiation, in 1984, individual records of Venezuela's 1980-round census were not available. Even if they were, it is doubtful an income variable could be obtained since doing so for 1971 was purely by chance. Correspondingly, an income variable is not found in 1980-round data that was available through CELADE (e.g., Ecuador's). For further discussion of CELADE data, see Centro Latinoamericano de Demografia (1974, 1978, 1984). With regard to the age twenty-one criterion, this reflects the centrality of educational attainment in the study's conceptual focus. By this time, most people would have completed schooling, whereas an age fifteen criterion, used by the Census for occupation and employment questions, might distort findings on education.

7. Third World censuses often under report labor force participation, partly due to the prevalence of unpaid, household-related, and/or informal employment; Phillips' (1987) study of Ecuador's rural Sierra provides an example. On this basis, Recchini de Lattes and Wainerman (1986), addressing the issue for females, find fault with most Latin American census statistics. They note, however, "Venezuela has carefully questioned women about their labor force participation... the level of precision... resembles a household survey, in which the activity status is measured through a series of six questions" (p. 745). Also relevant is Mohan's (1986: ch. 7) finding that census-based participation rates for Bogota were not systematically lower than those derived from World Bank surveys. These observations, one from a critical perspective, support using census data; but doing so with caution, keeping in mind that official statistics may under report some labor market segments. More generally, questioning census data accuracy (for the United States as well as Third World nations) is an important function, but such data remain among the best available. Hence, its use must continue, even as improvements take place.

8. This suggests the interesting possibility of gender differences in migration motivation. That is, males may move to more developed areas for wages, while females may be attracted by a greater likelihood of labor force participation. For an example of research considering gender differences in migration motivation, see Monk (1981) and Monk and Alexander (1986).

9. The preceding paragraph focusses on substantive characteristics of the equation system by which individual labor market experiences are examined. This footnote points to a methodological issue; namely, that aggregate measures of the dependent variables EDUC and WAGE are included in the place variables STRUCTURE (where they also play a strong role) and PRESSURE. On this basis, some may conclude there is a tautology; i.e., that the dependent variables are, in effect, being explained by identities. In defense of the methodology used here, two points should be noted. First, while it is true that individual educational attainment is likely to be greater in places with higher attainment in the aggregate, also of interest is the role of other personal attributes such as GENDER and AGE; in this regard, place characteristics associated with development may be seen as control variables. Second, the methodology indicates how place characteristics related to development alter the role of each personal attribute, i.e., their indirect effects on EDUC, LFP, and WAGE; and this is a salient cornerstone of the investigation.

10. Separate analyses for males and females were carried out for labor force participation and wages received. F-ratios indicate these regressions are significant at the 0.01 level or better.

11. Another noteworthy aspect of Venezuelan employment is that males

and females in manufacturing are approximately equal in percentage terms (Table 5.3). According to Perez-Sainz and Zarembka (1979), Venezuelan industrialization featured a coalition of interests which recognized industrial expansion, labor, and multi-class agendas. But they also note that Venezuela's economy was the most dynamic in Latin America in the 1960s, creating a continual need for labor in a nation with relatively low population. When this need subsides, will manufacturing employment become more of a male domain, as it did in Guatemala under similar circumstances (Chinchilla 1977)?

Chapter 6: Policy aspects of development and regional change I: Population movements from Ecuador's rural Sierra

1. In 1982 Ecuador had 126 cantones, but these were reconstituted to 1974 boundaries because analyses employ data from both years together. Also, the Galapagos Island province (with three cantones) was omitted from consideration because, in terms of issues addressed here, it represents a situation that is radically different from mainland Ecuador.

2. Description of these regions and related matters draws largely on Blakemore and Smith (1983: 253–324), Handelman (1980), Instituto Geografico Militar (1982), James and Minkel (1986: 320–33), Morris (1987: 228–41), Odell and Preston (1978), Schodt (1987), and Weil, Black, Blutstein, McMorris, Mersereau, Munson, and Parachini (1973).

3. Among urban areas in the Sierra are Quito, Ecuador's national capital and second largest city; Cuenca, its third largest city; and Ambato, fifth largest and a rapidly growing economic center.

4. The remainder of this chapter is a restatement of research originally reported in Brown, Brea, and Goetz (1988). Permission for use obtained from *Economic Geography*.

5. Circulation research has tended to focus on Africa and Southeast Asia (Goldstein 1978), but Latin America has received increasing attention in recent years. Circulation there has grown considerably because of structural changes related to land tenancy practices, commercialization of agriculture, mechanization, and other shifts leading to increased use of temporary labor and proletarianization of the agricultural work force (Balan 1983; Commander and Peek 1986; Dinerman 1982; Goodman and Redclift 1977; Matos Mar and Mejia 1982; Peek 1982; Roberts 1982, 1985). Recent publications on circulation, other than those focussing on Latin America or already cited, include Chapman and Prothero, (1985), Forbes (1981), Hugo (1982), and Prothero and Chapman (1985).

6. Among the few comparable examples, see Brea (1986); Costello, Leinbach, and Ulack (1987); and Hugo (1985).

7. At all times, arrimados provided an important source of cash income for the family and community.

8. Indeed, some hacendados, interested in modernizing their agricultural enterprise and moving away from the restrictions and costs of huasipungo, were among the early proponents of land reform. Areas in which this response was particularly prevalent include Mejia in the province of Pichincha and Salcedo in Cotopaxi (Programa Nacional de Regionalizacion Agraria 1979: Documento B, 111–20) (For locations, see Appendix 6.1).

9. Indicative of the significance of temporary migration among minifundistas, approximately 52 percent of their wage earnings in 1974 were from non-agricultural sources. Since wage employment of this sort is scarce in rural areas, a large portion must have taken place in cities (Commander and Peek 1986: 86).

10. For comparison, cantones excluded from the analysis because they represent major urban centers show the following MODERNITY scores: Quito in Pichincha, +3.67; Ambato in Tungurahua, +0.98; and Cuenca in Azuay, +1.48.

11. For a discussion of this CELADE data set, see Centro Latinoamericano de Demografia (1984).

12. In addition to criteria listed immediately below the demarcation of this note, n-sizes were reduced by software limitations. Records with missing data were automatically expunged, and restricted workspace required deleting another 10 percent of those still available.

13. Analysis only considered persons from the Sierra because, as noted earlier, land reform was most concerned with traditional agrarian systems found in that region. Delineating the rural Sierra was done by omitting cantones containing the large cities of Quito, Cuenca, and Ambato. Circulators were defined as persons whose canton of 'present' residence differs from that of 'usual' residence; migrants as persons with a difference in cantones of 'usual' and 'previous' residence. 'Present', 'usual', and 'previous' are terms used by the census.

14. The nine-year criterion is to ensure that migration occurred during or after 1974, the year represented by place characteristics giving rise to out-migration. A similar criterion is not relevant to circulation since it is short term.

15. OUTCIRC and OUTMIG are specified as 1 for a circulator or migrant, 0 otherwise; GENDER as 1 for male, 2 for female; AGE as years

since birth; MARITAL as 1 if single, 2 if widowed or divorced, 3 if married or in a union; EDUC as years of schools attended. TIME, PRODUCTION ORIENTATION, SIZE, and MODERNITY are standardized variables with values ranging, approximately, from -3.0 to +3.0 for the whole of Ecuador. Previous residence is canton of 'usual' residence for circulators, 'previous' for migrants, and 'present' for non-circulators/non-migrants.

16. For a discussion of logistic regression see Aldrich and Nelson (1984) or Wrigley (1976).

17. Standardized indices for out-circulation were derived by assigning a 1.0 value to the highest rate; computing the ratio of other out-circulation rates to the highest rate; then using that ratio as each canton's standardized index. A similar procedure was followed for out-migration.

18. Programa Nacional de Regionalizacion Agraria (1979) has been the primary source of information concerning ground-level conditions in cantones and their relationship with circulation and migration. Additional detail on specific rural Sierran communities is available in Belote and Belote (1985), Bromley (1985), Lentz (1985), Martinez (1985), Orbe and Chontasi (1985), Pachano (1985), and Rodas (1985). In that Peru presents a comparable situation, also of interest is Alberts (1983), Laite (1985), Skeldon (1985).

Chapter 7: Policy aspects of development and regional change II: The juxtaposition of national policies and local socioeconomic structures in Ecuador

1. This chapter draws on research originally reported in Lawson (1986).

2. Although this situation might be offset by 'points', additional money paid at the loan's inception, resources must be available to cover that amount. Small farms would continue, therefore, to be disadvantaged.

3. To indicate the extent of distortion in value, when Ecuador finally indexed the sucre to major world currencies, in 1983, a 63 percent devaluation was required (Sigma One 1985).

4. These effects are illustrated by the following ceteris paribus examples. Assuming a free market exchange rate of 100 sucres per dollar, imports of wheat valued at $10 per kilo on world markets would cost Ecuadorian consumers 1000 sucres. Overvaluing the sucre at 25 per dollar reduces that cost to 250 sucres. Hence, imported wheat is artificially cheapened relative to domestically produced wheat,

enjoying a price advantage that may strangle domestic production. Simultaneously, Ecuadorian exporters are disadvantaged by overvaluation, earning fewer sucres per dollar value of exports (Gibson 1971). Assume 800 sucres of production costs for a quantity of bananas selling on world markets for $10. At free market currency values (again, 100 sucres per dollar) exporters receive 1000 sucres, yielding a 200 sucre profit. If, however, the sucre is overvalued at 25 sucres per dollar, only 250 are received, resulting in a net loss. Again, then, production incentives are reduced and agricultural stagnation encouraged.

5. Variables chosen were limited by availability in Ecuador's 1974 Censuses of Agriculture and Population.

Chapter 8: Third World development as the local articulation of world economic and political conditions, donor-nation actions, and government policies: Concluding observations

1. The remainder of this chapter is a restatement of portions of research originally reported in Brown (1988). Permission for use obtained from *Economic Geography*.

2. These examples focus on specific external forces. A broader question is why several Asian countries classified as newly industrialized (NIC) have crossed a threshold into self-sustaining growth (e.g., Korea, Taiwan), while Latin American NICs have not (e.g., Venezuela, Argentina, Chile). Relevant to this are Browett (1985), Tang and Worley (1988), and Morawetz (1980). The latter compares Colombia with East Asian nations, particularly Hong Kong, in their efforts to manufacture clothes for export. Differences in success are traced to national policies and cultural factors affecting labor productivity, quality control, and the like.

3. For example, as reported on Saturday April 11, 1987 in the newspaper *Guayaquil Expresso*, in the year 1985–86 alone production grew from 6.59 to 9.87 tons, an increase of nearly fifty percent.

4. Neglect of place knowledge may be especially common in research concerned with orthodox conceptualizations of development. An example is Bradshaw (1988) finding an inverse relationship between urbanization and educational attainment changes in Kenya, leading to a claim that this counters modernization theory. A critical variable in his analysis is the proportion of persons of secondary school age who are in school. Linking this variable to modernization theory neglects the Third World reality that (in settings where six years of education is above the average) most secondary school-age persons are

drawn to urban areas for employment, rather than educational, opportunities. Another example is studies linking development with labor force shifts towards the manufacturing and service sectors (e.g., Casetti and Pandit 1987; Pandit 1986). Such shifts often are attributed to development forces associated with orthodox conceptualizations, without testing alternative hypotheses recognizing the preponderance, and likely importance, of policies and donor-nation actions that promote manufacturing and service activities. For a discussion of alternative hypotheses and a study that does recognize contextual variation in processes related to labor force shifts among economic sectors, see Pandit and Casetti (1989).

5. Gilbert's (1986: 544) commentary on Armstrong and McGee (1985) indicates he also sees verticalization is a broadly applicable force: "Armstrong's discussion of the regional elite of Cuenca... shows how one family's economic interests have gradually diversified and spread from agriculture, into urban activities and now beyond the region altogether. The description of this process is useful because it does reflect a widespread phenomenon in Latin America."

6. For more on this issue, see Smith's (1987) criticism of locality studies in the United Kingdom and Cooke's (1987) highly informative reply.

7. Care must be maintained, however, in that *similarly named* policies may *differ* in significant aspects. Hence, improving peasant welfare was a cornerstone of land reform in Mexico whereas more recent programs (e.g., in Costa Rica, Ecuador, Peru) tend to emphasize production increases (Chapters Three and Six; Peek and Standing 1982b). Price policy provides another illustration (Tolley, Thomas, and Wong 1982).

8. Resurrection of this traditional, but often neglected, forte of Geography owes much to research grounded in political economy thinking, an irony given tensions between it and other aspects of the discipline. Nevertheless, political economy's concern with place has been to illustrate or validate a general construct; whereas this book stresses place as an object of study in its own right, from which generalizations will emerge. For a general discussion on trends in regional geography, see Pudup (1988). Also of interest is the work of Robert S. Platt, a major influence on Taaffe. Exemplary is Platt's (1943) book on Latin America, which draws general conclusions after detailed, place-based analyses rather than establishing a broad framework at the outset as is common in current-day regional studies.

References

Adams, D.W., Graham, D.W., and von Pischke, J.D. (eds.) (1984) *Undermining Rural Development With Cheap Credit*, Boulder, West-view.

Adelman, I. and Morris, C.T. (1971) *Society, Politics, and Economic Development: A Quantitative Approach*, Baltimore, Johns Hopkins University Press.

Adelman, I. and Morris, C.T. (1973) *Economic Growth and Social Equity in Developing Countries*, Stanford, Stanford University Press.

Alberts, T. (1983) *Agrarian Reform and Rural Poverty: A Case Study of Peru*, Boulder, Westview.

Aldrich, J.H. and Nelson, F.D. (1984) *Linear Probability, Logit, and Probit Models*, Beverly Hills, Sage, Quantitative Applications in the Social Sciences.

Alonso, W. (1977) 'A theory of movements', in N.M. Hansen (ed.) *Human Settlement Systems: International Perspectives on Structure, Change, and Public Policy*, Cambridge, Ballinger, pp. 197–211.

Amin, S. (1977) 'Underdevelopment and dependence in Black Africa — origins and contemporary forms', in J. Abu-Lughod and R. Hay, Jr. (eds.) *Third World Urbanization*, Chicago, Maaroufa, pp. 140–50.

Anderson, D. (1982) *Small Industry in Developing Countries: Some Issues*, Washington, World Bank, Staff Working Paper 518.

Anderson, D. and Leiserson, M.W. (1980) 'Rural-nonfarm employment in developing countries', *Economic Development and Cultural Change* 28, 227–48.

Anselin, L. (1988) *Spatial Econometrics: Methods and Models*, Dordrecht, Martinus-Nijhoff.

Arizpe, L. (1977) 'Women in the informal sector: the case of Mexico City', *Signs* 3, 25–37.

Armstrong, W. and McGee, T.G. (1985) *Theatres of Accumulation: Studies in Asian and Latin American Urbanization*, London and New York, Methuen.

Arndt, H.W. (1987) *Economic Development: The History of an Idea*, Chicago, University of Chicago Press.

Bach, R.L. and Schraml, L.A. (1982) 'Migration, crisis, and theoretical conflict', *International Migration Review* 16, 320–41.

Balan, J. (1983) 'Agrarian structure and internal migration in a historical perspective: Latin American case studies', in P.A. Morrison (ed.) *Population Movements: Their Forms and Functions in Urbanization and Development*, Liege, Ordina Editions, pp. 151–85.

Barber, G.M. and Milne, W.J. (1988) 'Modelling internal migration in Kenya: an econometric analysis with limited data', *Environment and Planning A* 20, 1185–96.

Beals, R.E., Levy, M.B., and Moses, L.N. (1967) 'Rationality and migration in Ghana', *Review of Economics and Statistics* 49, 480–6.

Belote, J. and Belote, L. (1985) 'Vertical circulation in southern Ecuador', in R.M. Prothero and M. Chapman (eds.) *Circulation in Third World Countries*, London, Routledge and Kegan Paul, pp. 160–77.

Belsky, E. (1988) 'Regional development and potato marketing in Tungurahua Province, Ecuador', Project Report, International Development Program, Clark University.

Beneria, L. (ed.) (1982) *Women and Development: The Sexual Division of Labor in Rural Societies*, New York, Praeger.

Beneria, L. and Sen, G. (1981) 'Accumulation, reproduction, and women's role in economic development: Boserup revisited', *Signs* 7, 279–98.

Berry, B.J.L. (1961) 'Basic patterns of economic development', in N. Ginsburg, *Atlas of Economic Development*, Chicago, University of Chicago Press, pp. 110–19.

Berry, B.J.L. (1969) 'Relationships between regional economic development and the urban system: the case of Chile', *Tijdschrift voor Economishe en Sociale Geografie* 60, 283–307.

Berry, B.J.L. and Dahmann, D.C. (1977) 'Population redistribution in the United States in the 1970s', *Population and Development Review* 3, 443–71.

Berry, B.J.L. and Horton, F.E. (1970) *Geographic Perspectives on Urban Systems: With Integrated Readings*, Englewood Cliffs, Prentice-Hall.

Berry, B.J.L., Conkling, E.C., and Ray, D.M. (1987) *Economic Geography: Resource Use, Locational Choices, and Regional Specialization in the Global Economy*, Englewood Cliffs, Prentice-Hall.

Betancourt, J.F. (1978) 'Estimating interstate internal migration from place-of-birth data', *Revista Geografica* 88, 61–77.

Biggar, J.C. (1979) *The Sunning of America: Migration to the Sunbelt*, Washington, Population Reference Bureau, Population Bulletin 34–1.

Bilsborrow, R.E. (1987) 'Population pressures and agricultural development in developing countries: a conceptual framework and recent evidence', *World Development* 15, 183–203.

Bilsborrow, R.E., McDevitt, T.M., Kossoudji, S., and Fuller, R. (1987) 'The impact of origin community characteristics on rural–urban out-migration in a developing country', *Demography* 24, 191–210.

Blakemore, H. and Smith, C.T. (1983) *Latin America: Geographical Perspectives, Second Edition*, London and New York, Methuen.

Blankstein, C.A. and Zuvekas, C. (1973) 'Agrarian reform in Ecuador: an evaluation of past efforts and the development of a new approach', *Economic Development and Cultural Change* 22, 73–94.

Blaug, M. (1973) *Education and the Employment Problem in Developing Countries*, Geneva, International Labour Office.

Blaug, M. (1976) 'Human capital theory: a slightly jaundiced survey', *Journal of Economic Literature* 14, 827–56.

Blutstein, H.I., Andersen, L.C., Betters, E.C., Dombrowski, J.H., and Townsend, C. (1970) *Area Handbook for Costa Rica*, Washington, United States Government Printing Office.

Blutstein, H.I., Edwards, J.D., Johnston, K.T., McMorris, D.S., and Rudolph, J.D. (1977) *Area Handbook for Venezuela*, Washington, United States Government Printing Office.

Booth, D. (1985) 'Marxism and development sociology: interpreting the impasse', *World Development* 13, 761–87.

Boserup, E. (1970) *Women's Role in Economic Development*, New York, St. Martin's Press.

Boulding, E. (1980) *Women: The Fifth World*, New York, Foreign Policy Association.

Bourgin, F. (1989) *The Great Challenge: The Myth of Laissez-Faire in the Early Republic*, New York, George Braziller.

Bowman, M.J. (1980) 'Education and economic growth: an overview', in T. King (ed.) *Education and Income*, Washington, World Bank, Staff Working Paper 402, pp. 1–71.

Bradshaw, Y.W. (1985) 'Overurbanization and underdevelopment in Sub-Saharan Africa: a cross-national study', *Studies in Comparative International Development* 20, 74–101.

Bradshaw, Y.W. (1987) 'Urbanization and underdevelopment: a global study of modernization, urban bias, and economic dependency', *American Sociological Review* 52, 224–39.

Bradshaw, Y.W. (1988) 'Urbanization, personal income, and physical quality of life: the case of Kenya', *Studies in Comparative International Development* 23, 15–40.

Brea, J.A. (1986) *Effects of Structural Characteristics and Personal Attributes On Labor Mobility in Ecuador*, unpublished Ph.D. dissertation, Ohio State University, Department of Geography; also in Studies on the Interrelationships Between Migration and Development in Third World Settings, Discussion Paper 34, Department of Geography, Ohio State University.

Brodsky, D.A. and Rodrik, D. (1981) 'Indicators of development and data availability: the case of the PQLI', *World Development* 9, 695–9.

Bromley, R. (1985) 'Circulation within systems of periodic and daily markets: the case of central highland Ecuador', in R.M. Prothero and M. Chapman (eds.) *Circulation in Third World Countries*, London, Routledge and Kegan Paul, pp. 325–49.

Brookfield, H. (1975) *Interdependent Development*, London, Methuen.

Browder, J.O. (1987) 'Brazil's export promotion policy (1980–1984): impacts on the Amazon's industrial wood sector', *Journal of Developing Areas* 21, 285–304.

Browett, J. (1985) 'The newly industrializing countries and radical theories of development', *World Development* 13, 789–803.

Brown, L.A. (1981) *Innovation Diffusion: A New Perspective*, London and New York, Methuen.

Brown, L.A. (1988) 'Reflections on Third World development: ground level reality, exogenous forces, and conventional paradigms', *Economic Geography* 64, 255–78.

Brown, L.A. and Goetz, A.R. (1987) 'Development-related contextual effects and individual attributes in Third World migration processes: a Venezuelan example', *Demography* 24, 497–516.

Brown, L.A. and Holmes, J. (1971a) 'Intra-urban migrant lifelines: a spatial view', *Demography* 8, 103–22.

Brown, L.A. and Holmes, J. (1971b) 'Search behavior in an intra-urban migration context: a spatial perspective', *Environment and Planning*

A 3, 307–26.

Brown, L.A. and Jones, J.P. III (1985) 'Spatial variation in migration processes and development: a Costa Rican example of conventional modeling augmented by the expansion method', *Demography* 22, 327–52.

Brown, L.A. and Kodras, J.E. (1987) 'Migration, human resource transfers, and development contexts: a logit analysis of Venezuelan data', *Geographical Analysis* 19, 243–63.

Brown, L.A. and Lawson, V.A. (1985a) 'Migration in Third World settings, uneven development, and conventional modeling: a case study of Costa Rica', *Annals of the Association of American Geographers* 75, 29–47.

Brown, L.A. and Lawson, V.A. (1985b) 'Rural destined migration in Third World settings: a neglected phenomenon?', *Regional Studies* 19, 415–32.

Brown, L.A. and Lawson, V.A. (1989) 'Polarization reversal, migration-related shifts in human resource profiles, and spatial growth policies: a Venezuelan study', *International Regional Science Review* 12, 165–88.

Brown, L.A. and Moore, E.G. (1970) 'The intra-urban migration process: a perspective', *Geografiska Annaler, Series B* 52, 1–13; also in *Yearbook of the Society for General Systems Research* 15, 109–22.

Brown, L.A. and Sanders, R.L. (1981) 'Toward a development paradigm of migration: with particular reference to Third World settings', in G.F. DeJong, and R.W. Gardner (eds.) *Migration Decision Making: Multidisciplinary Approaches to Micro-Level Studies in Developed and Developing Countries*, New York, Pergamon Press, pp. 149–85.

Brown, L.A. and Stetzer, F.C. (1984) 'Development aspects of migration in Third World settings: a simulation, with implications for urbanization', *Environment and Planning A* 16, 1583–1603.

Brown, L.A., Brea, J.A., and Goetz, A.R. (1988) 'Policy aspects of development and individual mobility: migration and circulation from Ecuador's rural Sierra', *Economic Geography* 64, 147–70.

Brown, L.A., England, K.V.L., and Goetz, A.R. (1989) 'Location, social categories, and individual labor market experiences in developing economies: the Venezuelan case', *International Regional Science Review* 12, 1–28.

Buvinic, M., Lycette, M.A., and McGreevey, W.P. (eds.) (1983) *Women and Poverty in the Third World*, Baltimore, Johns Hopkins University Press.

Caporaso, J.A. (ed.) (1987) *A Changing International Division of Labor*, London, Frances Pinter, Boulder, Lynne Rienner.

Caporaso, J.A. and Zare, B. (1981) 'An interpretation and evaluation of dependency theory', in H. Munoz (ed.) *From Dependency to Development: Strategies to Overcome Underdevelopment and Inequality*, Boulder, Westview, pp. 43–56.

Cardoso, F.H. and Faletto, E. (1979) *Dependency and Development in Latin America*, Berkeley, University of California Press.

Carvajal, M.J. and Geithman, D.T. (1974) 'An economic analysis of migration in Costa Rica', *Economic Development and Cultural Change* 23, 105–22.

Casetti, E. (1964) 'Multiple discriminant functions', Technical Report 11, ONR Task 389–135, Contract Nonr 1228(26), Office of Naval Research, Geography Branch.

Casetti, E. (1965) 'Classificatory and regional analysis by discriminant iterations', Technical Report 12, ONR Task 389–135, Contract Nonr 1228(26), Office of Naval Research, Geography Branch.

Casetti, E. (1972) 'Generating models by the expansion method: applications to geographical research', *Geographical Analysis* 4, 81–91.

Casetti, E. (1982) 'Mathematical modeling and the expansion method', in R.B. Mandal (ed.) *Statistics for Geographers and Social Scientists*, New Delhi, Concept Publishing, pp. 81–95.

Casetti, E. (1986) 'The dual expansion method: an application for evaluating the effects of population growth on development', *IEEE (Institute of Electrical and Electronics Engineers) Transactions on Systems, Man, and Cybernetics* 16, 29–39.

Casetti, E. and Jones, J.P. III (1987) 'Spatial aspects of the productivity slowdown: an analysis of U.S. manufacturing data', *Annals of the Association of American Geographers* 77, 76–88.

Casetti, E. and Pandit, K. (1987) 'The non-linear dynamics of sectoral shifts', *Economic Geography* 63, 241–58.

Center for International Research (1985) *Women of the World: A Chartbook for Developing Regions*, Washington, United States Bureau of the Census, Center for International Research.

Centro Economico Para America Latina (1983) *Five Studies on the Situation of Women in Latin America*, Santiago, United Nations, Estudios e Informes de la CEPAL (Centro Economico Para America Latina).

Centro Latinoamericano de Demografia (1974) *Boletin del Banco de Datos*, Santiago, United Nations, Centro Latinoamericano de Demografia.

Centro Latinoamericano de Demografia (1978) *The CELADE Data Bank: Data Available and Procedures to Obtain Tapes and Tabulations*, Santiago, United Nations, Centro Latinoamericano de Demografia.

Centro Latinoamericano de Demografia (1984) *Boletin del Banco de Datos*, Santiago, United Nations, Centro Latinoamericano de Demografia.

Chapman, M. and Prothero, R.M. (eds.) (1985) *Circulation in Population Movement: Substance and Concepts from the Melanesian Case*, London, Routledge and Kegan Paul.

Chaves, L.F. (1973) 'The economic base and functional structure of Venezuelan cities', Polish Academy of Sciences, research report.

Chen, C.Y. (1978) *Desarrollo Regional–Urbano y Ordenamiento del Territorio: Mito y Realidad*, Caracas, Universidad Catolica Andres Bello.

Chen, C.Y. and Picouet, M. (1979) *Dinamica de la Poblacion: Caso de Venezuela*, Caracas, Universidad Catolica Andres Bello, Instituto de

Investigaciones Economicas y Sociales.

Chenery, H. (1979) *Structural Change and Development Policy*, Baltimore, Johns Hopkins University Press.

Chenery, H. (1984) 'Policy instruments and development alternatives', in G.M. Meier (ed.) *Leading Issues in Economic Development, Fourth Edition*, Oxford and New York, Oxford University Press, pp. 732–43.

Chenery, H. and Syrquin, M. (1975) *Patterns of Development: 1950-1970*, Oxford and New York, Oxford University Press.

Chinchilla, N.S. (1977) 'Industrialization, monopoly capitalism, and women's work in Guatemala', *Signs* 3, 38–56.

Chorley, R.J. (1962) 'Geomorphology and general systems theory', Geological Survey Professional Paper 500–B, United States Government Printing Office.

Cleaver, K.M. (1985) *The Impact of Price and Exchange Rate Policies on Agriculture in Sub-Saharan Africa*, Washington, World Bank, Staff Working Paper 728.

Cochrane, S.H. (1979) *Fertility and Education: What Do We Really Know?*, Baltimore, Johns Hopkins University Press.

Colclough, C. (1982) 'The impact of primary schooling on economic development: a review of the evidence', *World Development* 10, 167–85.

Commander, S. and Peek, P. (1986) 'Oil exports, agrarian change and the rural labour process: the Ecuadorian Sierra in the 1970s', *World Development* 14, 79–96.

Conaway, M.E. (1976) *Still Guahibo, Still Moving: A Study of Circular Migration and Marginality in Venezuela*, unpublished Ph.D. dissertation, University of Pittsburgh, Department of Anthropology.

Conaway, M.E. (1977) 'Circular migration in Venezuelan frontier areas', *International Migration* 15, 35–42.

Connell, J., Dasgupta, B., Laishley, R. and Lipton, M. (1976) *Migration from Rural Areas: Evidence from Village Studies*, Oxford, Oxford University Press.

Cooke, P. (1987) 'Clinical inference and geographic theory', *Antipode* 19, 69–78.

Cortes, M., Berry, A., and Ishaq, A. (1987) *Success in Small and Medium-Scale Enterprises: The Evidence from Colombia*, Oxford and New York, Oxford University Press.

Costello, M.A., Leinbach, T.R., and Ulack, R. (1987) *Mobility and Employment in Urban Southeast Asia: Examples from Indonesia and the Philippines*, Boulder, Westview.

Dinerman, I.R. (1982) *Migrants and Stay-at-Homes: A Comparative Study of Rural Migration from Michoacan, Mexico*, La Jolla, University of California at San Diego, Center for US–Mexican Studies, Monograph Series 5.

Dore, E. (1988) *The Peruvian Mining Industry: Growth, Stagnation, and Crisis*, Boulder, Westview.

Dos Santos, T. (1974) 'Brazil: the origins of a crisis', in R.H. Chilcote and J.C. Edelstein (eds.) *Latin America: The Struggle With Dependen-*

cy and Beyond, New York, John Wiley, Schenkman Publishing, pp. 409–90.

Draper, E. (1985) 'Women's work and development in Latin America', *Studies in Comparative International Development* 20, 3–30.

Economic Perspectives (1985) 'Grain policy in Ecuador', technical report, Ministerio de Agricultura y Ganaderia, Quito.

Ettema, W.A. (1979) 'Geographers and development', *Tijdschrift voor Economishe en Sociale Geografie* 70, 66–74.

Ewell, J. (1984) *Venezuela: A Century of Change*, Stanford, Stanford University Press.

Fair, T.J.D. (1982) *South Africa: Spatial Frameworks for Development*, Capetown, Juta.

Falaris, E.M. (1979) 'The determinants of internal migration in Peru: an economic analysis', *Economic Development and Cultural Change* 27, 327–41.

Fei, J.C.H. and Ranis, G. (1964) *Development of the Labor Surplus Economy: Theory and Policy*, Homewood, Irwin.

Fernandez-Kelly, M.P. (1981) 'Development and the sexual division of labor: an introduction', *Signs* 7, 268–78.

Fields, G.S. (1979) 'Lifetime migration in Colombia: tests of the expected income hypothesis', *Population and Development Review* 5, 247–65.

Fields, G.S. (1980) *Poverty, Inequality, and Development*, Cambridge, Cambridge University Press.

Fields, G.S. (1982) 'Place-to-place migration in Colombia', *Economic Development and Cultural Change* 30, 539–58.

Fields, G.S. and Schultz, T.P. (1980) 'Regional inequality and other sources of income variation in Colombia', *Economic Development and Cultural Change* 28, 447–67.

Findley, S.E. (1977) *Planning for Internal Migration: A Review of Issues and Policies in Developing Countries*, Washington, United States Government Printing Office.

Findley, S.E. (1981) 'Rural development programmes: planned versus actual migration outcomes', in G.J. Demko and R.J. Fuchs (eds.) *Population Distribution Policies in Development Planning*, New York, United Nations, Department of International Economic and Social Affairs, Population Studies 75, pp. 144–66.

Findley, S.E. (1982) 'Methods of linking community-level variables with migration survey data', in United Nations Economic and Social Commission for Asia and the Pacific, *National Migration Surveys: X. Guidelines for Analyses*. New York, United Nations, pp. 276–311.

Findley, S.E. (1987a) 'An interactive contextual model of migration in Ilocos Norte, the Philippines', *Demography* 24, 163–90.

Findley, S.E. (1987b) *Rural Development and Migration: A Study of Family Choices in the Philippines*, Boulder, Westview.

Findley, S.E., Gundlach, J., Kent, D.P., and Rhoda, R. (1979) *Rural Development, Migration, and Fertility: What Do We Know?*, Washington, United States Agency for International Development, Office of

Rural Development and Development Administration, Rural Development and Fertility Project, Final Report.

Flammang, R.A. (1979) 'Economic growth and economic development: counterparts or competitors?', *Economic Development and Cultural Change* 28, 47–61.

Forbes, D.K. (1981) 'Mobility and uneven development in Indonesia: a critique of explanations of migration and circular migration', in G.W. Jones and H.V. Richter (eds.) *Population Mobility and Development: Southeast Asia and the Pacific*, Canberra, Australian National University, Development Studies Centre, Monograph 27, pp. 51–70.

Forbes, D.K. (1984) *The Geography of Underdevelopment: A Critical Survey*, Baltimore, Johns Hopkins University Press.

Frank, A.G. (1967) *Capitalism and Underdevelopment in Latin America: Historical Studies of Chile and Brazil*, New York, Monthly Review Press.

Frank, A.G. (1986) 'The development of underdevelopment', in P.F. Klaren and T.J. Bossert (eds.) *Promise of Development: Theories of Change in Latin America*, Boulder, Westview, pp. 111–23.

Franklin, D.L. and Penn, J.B. (1985) 'Review of prices for rice, maize, and soybeans: summer season, July–September 1985', technical report, United States Agency for International Development, Ecuador Mission, Quito.

Freeman, D.B. and Norcliffe, G.B. (1985) *Rural Enterprise in Kenya: Development and Spatial Organization of the Nonfarm Sector*, Chicago, University of Chicago Department of Geography, Research Paper 214.

Friedmann, J. (1966) *Regional Development Policy: A Case Study of Venezuela*, Cambridge, MIT Press.

Friedmann, J. (1972) 'A general theory of polarized development', in N.M. Hansen (ed.) *Growth Centers In Regional Economic Development*, New York, Free Press, pp. 82–107.

Friedmann, J. (1973) *Urbanization, Planning, and National Development*, Beverly Hills, Sage.

Friedmann, J. (1975) 'The spatial organization of power in the development of urban systems', in J. Friedmann and W. Alonso (eds.) *Regional Policy: Readings in Theory and Application*, Cambridge, MIT Press, pp. 266–304.

Friedmann, J. and Wulff, R. (1976) 'The urban transition: comparative studies of newly industrializing societies', in C. Board, R.J. Chorley, P. Haggett, and D.R. Stoddart (eds.) *Progress in Geography, Volume 8*, New York, St. Martin's Press, pp. 1–93.

Frobel, F., Heinrichs, J. and Kreye, O. (1980) *The New International Division of Labor*, Cambridge, Cambridge University Press.

Fuchs, R.J. and Demko, G.J. (1978) 'The postwar mobility transition in Eastern Europe', *Geographical Review* 68, 171–82.

Furtado, C. (1976) *Economic Development of Latin America, Second Edition*, Cambridge, Cambridge University Press.

Gaile, G.I. (1980) 'The spread–backwash concept', *Regional Studies* 14,

15–25.

Gardner, R.W. (1981) 'Macrolevel influences on the migration decision process', in G.F. DeJong and R.W. Gardner (eds.) *Migration Decision Making: Multidisciplinary Approaches to Micro-level Studies in Developed and Developing Countries*, New York, Pergamon Press, pp. 59–89.

Ghosh, P.K. (ed.) (1984) *Third World Development: A Basic Needs Approach*, Westport, Greenwood Press.

Gibson, C.R. (1971) *Foreign Trade in the Economic Development of Small Nations: The Case of Ecuador*, New York, Praeger.

Gilbert, A. (1986) 'Latin American studies', *Progress in Human Geography* 10, 541–52.

Ginsburg, N. (1961) *Atlas of Economic Development*, Chicago, University of Chicago Press.

Glantz, F.B. (1973) *The Determinants of the Interregional Migration of the Economically Disadvantaged*, Boston, Federal Reserve Bank of Boston, Research Report Series; also unpublished Ph.D. dissertation, Department of Economics, Syracuse University.

Gober-Meyers, P. (1978a) 'Employment motivated migration and economic growth in post-industrial market economies', *Progress in Human Geography* 2, 207–29.

Gober-Meyers, P. (1978b) 'Interstate migration and economic growth: a simultaneous equations approach', *Environment and Planning A* 10, 1241–52.

Godfrey, E.M. (1973) 'Economic variables and rural urban migration: some thoughts on the Todaro hypothesis', *Journal of Development Studies* 10, 66–78.

Goldscheider, C. (1987) 'Migration and social structure: analytic issues and comparative perspectives in developing nations', *Sociological Forum* 2, 674–96.

Goldstein, S. (1978) *Circulation in the Context of Total Mobility in Southeast Asia*, Honolulu, East–West Center, Paper 53 of the Population Institute.

Goldstein, S. (1981) 'Research priorities and data needs for establishing and evaluating population redistribution policies', in G.J. Demko and R.J. Fuchs (eds.) *Population Distribution Policies in Development Planning*, New York, United Nations, Department of International Economic and Social Affairs, Population Studies 75, pp. 183–203.

Golledge, R.G. (1988) 'Science and humanism in geography: multiple languages in multiple realities', in R.G. Golledge, H. Couclelis, and P. Gould (eds.) *A Ground for Common Search*, Santa Barbara, Santa Barbara Geographical Press, pp. 63–71.

Goodman, D. and Redclift, M. (1977) 'The boias-frias: rural proletarization and urban marginality in Brazil', *International Journal of Urban and Regional Research* 1, 348–64.

Gordon, D.L. (1978) *Employment and Development of Small Enterprises*, Washington, World Bank, Sector Policy Paper.

Gore, C. (1984) *Regions in Question: Space, Development Theory, and Regional Policy*, London and New York, Methuen.

Gotsch, C.H. (1972) 'Technical change and the distribution of income in rural areas', *American Journal of Agricultural Economics* 54, 326–41.

Gottschang, T.R. (1987) 'Economic change, disasters, and migration: the historical case of Manchuria', *Economic Development and Cultural Change* 35, 461–90.

Gould, P.R. (1976) 'Tanzania 1920–63: the spatial impress of the modernization process', in C.G. Knight and J.L. Newman (eds.) *Contemporary Africa: Geography and Change*, Englewood Cliffs, Prentice-Hall, pp. 423–37.

Greenwood, M.J. (1969) 'The determinants of labor migration in Egypt', *Journal of Regional Science* 9, 283–90.

Greenwood, M.J. (1971a) 'An analysis of the determinants of internal labor mobility in India', *Annals of Regional Science* 1, 137–51.

Greenwood, M.J. (1971b) 'A regression analysis of migration to urban areas of a less developed country: the case of India', *Journal of Regional Science* 11, 253–62.

Greenwood, M.J. (1975a) 'Research on internal migration in the United States: a survey', *Journal of Economic Literature* 8, 397–433.

Greenwood, M.J. (1975b) 'Simultaneity bias in migration models: an examination', *Demography* 12, 519–36.

Greenwood, M.J. (1978) 'An econometric model of internal migration and regional economic growth in Mexico', *Journal of Regional Science* 18, 17–31.

Greenwood, M.J., Ladman, J.R., and Siegel, B.S. (1981) 'Long term trends in migratory behavior in a developing country: the case of Mexico', *Demography* 18, 369–88.

Gregory, J.W. and Piche, V. (1981) 'The demographic process of peripheral capitalism illustrated with African examples', Working Paper 29, Centre for Developing-Area Studies, McGill University.

Griffin, K.B. (1976) 'Systems of labour control and rural poverty in Ecuador', in K.B. Griffin (ed.) *Land Concentration and Rural Poverty*, New York, Holmes and Meier, pp. 172–220.

Grunig, J.E. (1971) 'Communication and the economic decision-making processes of Colombian peasants', *Economic Development and Cultural Change* 19, 580–97.

Gwynne, R.N. (1986) *Industrialization and Urbanization in Latin America*, Baltimore, Johns Hopkins University Press.

Hagen, E.E. (1962) *On the Theory of Social Change: How Economic Growth Begins*, Homewood, Dorsey Press.

Hagerstrand, T. (1957) 'Migration and area', in D. Hannerberg, T. Hagerstrand, and B. Odeving (eds.) *Migration in Sweden: A Symposium*, Lund, Gleerup, Lund Studies in Geography 13, pp. 27–158.

Hall, C. (1985) *Costa Rica: A Geographical Interpretation in Historical Perspective*, Boulder, Westview, Dellplain Latin American Studies 17.

Handelman, H. (1980) 'Ecuadorian agrarian reform: the politics of limited change', *American Universities Field Staff Reports* 49, 1–19.

Hansen, N.M. (1971) *Intermediate Size Cities as Growth Centers: Applications for Kentucky, The Piedmont Crescent, The Ozarks, and*

Texas, New York, Praeger.

Hansen, N.M. (ed.) (1972) *Growth Centers in Regional Economic Development*, New York, Free Press.

Havens, A.E. and Flinn, W.L. (1975) 'Green revolution technology and community development: the limits of action programs', *Economic Development and Cultural Change* 23, 469–81.

Heilbroner, R.L. (1967) 'Counterrevolutionary America', *Commentary* 43 (4), 31–8.

Helmsing, A.H.J. (1986) *Firms, Farms, and the State in Colombia: A Study of Rural, Urban, and Regional Dimensions of Change*, London and Boston, Allen and Unwin.

Hicks, N. and Streeten, P. (1979) 'Indicators of development: the search for a basic needs yardstick', *World Development* 7, 567–80.

Hirschman, A.O. (1958) *The Strategy of Economic Development*, New Haven, Yale University Press; reprinted by Norton, New York, 1978.

Hirschman, A.O. (1984) 'A dissenter's confession: "The strategy of economic development" revisited', in G.M. Meier and D. Seers (eds.) *Pioneers in Development*, Oxford and New York, Oxford University Press, pp. 87–111.

Ho, S.P.S. (1980) *Small Scale Enterprise in Korea and Taiwan*, Washington, World Bank, Staff Working Paper 384.

Holmes, J. (1983) 'Industrial reorganization, capital restructuring, and locational change: an analysis of the Canadian automobile industry in the 1960s', *Economic Geography* 59, 251–71.

House, W.J. and Rempel, H. (1980) 'The determinants of interregional migration in Kenya', *World Development* 8, 25–35.

Hugo, G.J. (1982) 'Circular migration in Indonesia', *Population and Development Review* 8, 59–83.

Hugo, G.J. (1985) 'Structural change and labour mobility in rural Java', in G. Standing (ed.) *Labour Circulation and the Labour Process*, London, Croom Helm, pp. 46–88.

Hurtado, O. (1985) *Political Power in Ecuador*, Boulder, Westview.

Instituto Geografico Militar (1982) *Atlas del Ecuador*, Quito, Banco Central del Ecuador.

Instituto Nacional de Estadistica y Censos (1983) 'Sistema estadistica agropecuario nacional por muestras por areas', technical report.

International Center for Research on Women (1981) 'Women workers in Latin America: a structural analysis', International Center for Research on Women, Washington.

Isard, W. (ed.) (1960) *Methods of Regional Analysis: An Introduction to Regional Science*, Cambridge, MIT Press.

Jackson, R.H. and Hudman, L.E. (1986) *World Regional Geography: Issues for Today, Second Edition*, New York, John Wiley.

James, P.E. and Minkel, C.W. (1986) *Latin America, Fifth Edition*, New York, John Wiley.

Jelin, E. (1977) 'Migration and labor force participation of Latin American women: the domestic servants in the cities', *Signs* 3, 129–41.

Jelin, E. (1982) 'Women and the urban labour market', in R. Anker, M. Buvinic, and N.H. Youssef (eds.) *Women's Roles and Population Trends in the Third World*, London, Croom Helm, pp. 239–67.

Johnston, R.J. (1982) *The American Urban System: A Geographical Perspective*, New York, St. Martin's Press.

Johnston, R.J. (1984) *City and Society: An Outline for Urban Geography*, London, Hutchinson.

Johnston, R.J. (1987) *Geography and Geographers: Anglo American Human Geography Since 1945, Third Edition*, London, Edward Arnold.

Jonas, A. (1988) 'A new regional geography of localities', *Area* 20, 1–10.

Jones, J.P. III (1983) 'Parameter variation via the expansion method with tests for autocorrelation', *Modeling and Simulation* 14, 853–7.

Jones, J.P. III (1984a) 'A spatially-varying parameters model of AFDC participation: empirical analysis using the expansion method', *Professional Geographer* 36, 455–61.

Jones, J.P. III (1984b) *Spatial Parameter Variation in Models of AFDC Participation: Analyses Using the Expansion Method*, unpublished Ph.D. dissertation, Ohio State University, Department of Geography.

Jones, R.C. (1982) 'Regional income inequalities and government investment in Venezuela', *Journal of Developing Areas* 16, 373–89.

Jones, R.C. and Brown, L.A. (1985) 'Cross-national tests of a Third World development–migration paradigm, with particular attention to Venezuela', *Socio-Economic Planning Sciences* 19, 357–61.

Joseph, G.M. (1986) *Rediscovering the Past at Mexico's Periphery: Essays on the History of Modern Yucatan*, Tuscaloosa, University of Alabama Press.

Kahl, J.A. (1976) *Modernization, Exploitation, and Dependency in Latin America: Germani, Gonzalez Casanova, and Cardoso*, New Brunswick, Transaction Books.

Kau, J.B. and Sirmans, C.F. (1979) 'A recursive model of the spatial allocation of migrants', *Journal of Regional Science* 19, 47–56.

Kay, C. (1982) 'Agrarian change and migration in Chile', in P. Peek and G. Standing (eds.) *State Policies and Migration: Studies in Latin America and the Caribbean*, London, Croom Helm, pp. 35–79.

Kay, C. (1989) *Latin American Theories of Development and Underdevelopment*, London and New York, Routledge.

King, J. 1978. 'Interstate migration in Mexico', *Economic Development and Cultural Change* 27, 83–101.

Knight, C.G. and J.L. Newman (eds.) (1976) *Contemporary Africa: Geography and Change*, Englewood Cliffs, Prentice-Hall.

Laite, J. (1985) 'Circulatory migration and social differentiation in the Andes', in G. Standing (ed.) *Labour Circulation and the Labour Process*, London, Croom Helm, pp. 89–119.

Larson, D.A. and Wilford, W.T. (1979) 'The physical quality of life index: a useful social indicator', *World Development* 7, 581–84.

Lawson, V.A. (1986) *National Economic Policies, Local Variations in*

Structure of Production, and Uneven Regional Development: The Case of Ecuador, unpublished Ph.D. dissertation, Ohio State University, Department of Geography; also in Studies on the Interrelationships Between Development and Migration in Third World Settings, Discussion Paper 33, Department of Geography, Ohio State University.

Lawson, V.A. (1988) 'Government policy biases and Ecuadorian agricultural change', *Annals of the Association of American Geographers* 78, 433–52.

Lawson, V.A. and Brown, L.A. (1987) 'Structural tension, migration, and development: a case study of Venezuela', *Professional Geographer* 39, 179–88.

Layard, R. and Psacharopoulos, G. (1974) 'The screening hypothesis and returns to education', *Journal of Political Economy* 82, 985–98.

Ledent, J. (1982) 'Rural-urban migration, urbanization, and economic development', *Economic Development and Cultural Change* 30, 507–38.

Lee, S-H. (1985) *Why People Intend to Move: Individual and Community-Level Factors of Out-Migration in the Philippines*, Boulder, Westview.

Lentnek, B. (1980) 'Regional development and urbanization in Latin America: the relationship of national policy to spatial strategies', in R.N. Thomas and J.M. Hunter (eds.) *Internal Migration Systems in the Developing World, With Special Reference to Latin America*, Cambridge, Schenkman Publishing, pp. 82–113.

Lentz, C. (1985) 'Estrategias de reproduccion y migracion temporaria: indigenas de Cajabamba Chimborazo', *Ecuador Debate* 8, 194–215.

Levy, M. and Wadycki, W. (1972) 'Lifetime versus one year migration in Venezuela', *Journal of Regional Science* 12, 407–15.

Levy, M. and Wadycki, W. (1973) 'The influence of family and friends upon geographic labor mobility: an international comparison', *Review of Economics and Statistics* 55, 198–203.

Levy, M. and Wadycki, W. (1974a) 'Education and the decision to migrate: an econometric analysis of migration in Venezuela', *Econometrica* 42, 377–88.

Levy, M. and Wadycki, W. (1974b) 'What is the opportunity cost of moving? reconsideration of the effects of distance upon migration', *Economic Development and Cultural Change* 43, 198–214.

Lewis, O. (1961) *The Children of Sanchez: Autobiography of a Mexican Family*, New York, Random House.

Lewis, S. (1922) *Babbit*, New York, Harcourt Brace.

Lewis, W.A. (1954) 'Economic development with unlimited supplies of labor', *Journal of the Manchester School of Economics and Social Studies* 20, 139–92.

Lipton, M. (1976) *Why Poor People Stay Poor: A Study of Urban Bias in World Development*, Cambridge, Harvard University Press.

Lipton, M. (1980) 'Migration from rural areas of poor countries: the impact on rural productivity and income distribution', *World Development* 8, 1–24; also in R.H. Sabot (ed.) *Migration and the Labor Market in Developing Countries*, Boulder, Westview, pp. 191–228.

Lisk, F. (1977) 'Conventional development strategies and basic needs fulfillment: a reassessment of objectives and policies', *International Labour Review* 115, 175–91.

Little, I.M.D., Mazumdar, D., and Page, J.M. Jr. (1987) *Small Manufacturing Enterprises: A Comparative Analysis of India and Other Economies*, Oxford and New York, Oxford University Press.

Lockheed, M.E., Jamison, D.T., and Lau, L.J. (1980) 'Farmer education and farm efficiency: a survey', *Economic Development and Cultural Change* 29, 37–76.

Lowry, I.A. (1966) *Migration and Metropolitan Growth: Two Analytical Models*, San Francisco, Chandler.

Mabogunje, A.L. (1968) *Urbanization in Nigeria*, New York, Africana Publishing.

Mabogunje, A.L. (1970) 'Systems approach to a theory of rural-urban migration', *Geographical Analysis* 2, 1–17.

Mabogunje, A.L. (1981) *The Development Process: A Spatial Perspective*, New York, Holmes and Meier.

MacDonald, J. (1980) 'Planning implementation and social policy: an evaluation of Ciudad Guyana, 1965 and 1975', *Progress in Planning* 11, 7–211.

Mak, J. and Walton, G.M. (1973) 'The persistence of old technologies: the case of flatboats', *Journal of Economic History* 33, 444–51.

Markusen, A.R. (1985) *Profit Cycles, Oligopoly, and Regional Development*, Cambridge, MIT Press.

Markusen, A.R. (1987) *Regions: The Economics and Politics of Territory*, Totowa, Rowman and Littlefield.

Martinez, L. (1985) 'Migracion y cambios en las estrategias familiares de las communidades indigenas de la Sierra', *Ecuador Debate* 8, 110–28.

Maslow, A.H. (1968) *Toward a Psychology of Being, Second Edition*, Princeton, D. Van Nostrand.

Massey, D. (1983) 'Industrial restructuring as class restructuring: production decentralization and local uniqueness', *Regional Studies* 17, 73–89.

Massey, D. (1984) *Spatial Divisions of Labor: Social Structures and the Geography of Production*, London and New York, Methuen.

Mather, P.M. (1975) *Computational Methods of Multivariate Analysis in Physical Geography*, London, John Wiley.

Matos Mar, J. and Mejia, J.M. (1982) 'Casual work, seasonal migration, and agrarian reform in Peru', in P. Peek and G. Standings (eds.) *State Policies and Migration: Studies in Latin America and the Caribbean*, London, Croom Helm, pp. 81–119.

McClelland, D.C. (1961) *The Achieving Society*, Princeton, Van Nostrand.

McNulty, M.L. (1969) 'Urban structure and development: the urban system of Ghana', *Journal of Developing Areas* 3, 159–76.

McNulty, M.L. (1976) 'West African urbanization', in B.J.L. Berry (ed.) *Urbanization and Counter-Urbanization*, Beverly Hills, Sage, pp. 213–32.

Meier, G.M. (1984) *Leading Issues in Economic Development, Fourth Edition*, Oxford and New York, Oxford University Press.

Meier, G.M. and Seers, D. (1984) *Pioneers in Development*, Oxford and New York, Oxford University Press.

Mohan, R. (1986) *Work, Wages, and Welfare in a Developing Metropolis: Consequences of Growth in Bogota, Colombia*, Oxford and New York, Oxford University Press.

Monk, J. (1981) 'Social change and sexual differences in Puerto Rican rural migration', in O.H. Horst (ed.) *Papers In Latin American Geography in Honor of Lucia C. Harrison*, Muncie, Conference of Latin Americanist Geographers, pp. 28–43.

Monk J. and Alexander, C.S. (1986) 'Free port fallout: gender, employment and migration on Magarita Island, Venezuela', *Annals of Tourism Research* 13, 393–413.

Morawetz, D. (1977) *Twenty-Five Years of Economic Development: 1950 to 1975*, Baltimore, Johns Hopkins University Press.

Morawetz, D. (1980) *Why the Emperor's New Clothes are not Made in Colombia*, Washington, World Bank, Staff Working Paper 368.

Morris, A.S. (1987) *South America, Third Edition*, Totowa, Barnes and Noble.

Morris, M.D. (1979) *Measuring the Condition of the World's Poor: The Physical Quality of Life Index*, New York, Pergamon.

Morrison, P.A. (ed.) (1983) *Population Movements: Their Forms and Functions in Urbanization and Development*, Liege, Ordina Editions.

Myrdal, G. (1957) *Economic Theory and Underdeveloped Regions*, London, Duckworth.

Newsweek Staff (1983) 'The American dream: special anniversary issue', *Newsweek*, Spring.

Noor, A. (1981) *Education and Basic Human Needs*, Washington, World Bank, Staff Working Paper 450.

O'Brien, P.J. (1975) 'A critique of Latin American theories of dependency', in I. Oxaal, T. Barnett, and D. Booth (eds.) *Beyond the Sociology of Development: Economy and Society in Latin America and Africa*, London, Routledge and Kegan Paul, pp. 7–27.

Odell, P.R. and Preston, D.A. (1978) *Economies and Societies in Latin America: A Geographical Interpretation, Second Edition*, New York, John Wiley.

Okun, B. (1968) 'Interstate population migration and state income inequalities: a simultaneous equations approach', *Economic Development and Cultural Change* 16, 297–311.

Orbe, C. and Chontasi, L. (1985) 'Communidad, migracion, y capital: el caso Tabacundo', *Ecuador Debate* 8, 216–26.

Pachano, S. (1985) 'Migracion desde un pueblo serrano: Guaytacama', *Ecuador Debate* 8, 129–52.

Pack, H. (1987) *Productivity, Technology, and Industrial Development: A Case Study in Textiles*, Oxford and New York, Oxford University Press.

Page, J.M. Jr. (1979) *Small Enterprises in African Development: A*

Survey, Washington, World Bank, Staff Working Paper 363.

Page, J.M. Jr. and Steel, W.F. (1984) *Small Enterprise Development: Economic Issues from African Experience*, Washington, World Bank, Technical Paper 26.

Pandit, K. (1986) 'Sectoral allocation of labor force with development and the effect of trade activity', *Economic Geography* 62, 144–54.

Pandit, K. and Casetti, E. (1989) 'The shifting patterns of sectoral labor allocation during development: developed versus developing countries', *Annals of the Association of American Geographers* 79, 329–44.

Paris, J.D. (1970) 'Regional/structural analysis of population changes', *Regional Studies* 4, 425–43.

Peattie, L.R. (1980) 'Anthropological perspectives on the concepts of dualism, the informal sector, and marginality in developing urban economies', *International Regional Science Review* 5, 1–31.

Pedersen, P.O. (1975) *Urban–Regional Development in South America: A Process of Diffusion and Integration*, The Hague, Mouton.

Peek, P. (1979) 'Urban poverty, land reform, and rural emigration in Ecuador', Working Paper Series, World Employment Programme, International Labour Organization, Geneva.

Peek, P. (1980) 'Agrarian change and labour migration in the Sierra of Ecuador', *International Labour Review* 119, 609–21.

Peek, P. (1982) 'Agrarian change and labour migration in the Sierra of Ecuador', in P. Peek and G. Standing (eds.) *State Policies and Migration: Studies in Latin America and the Caribbean*, London, Croom Helm, pp. 121–45.

Peek, P. and Standing, G. (1979) 'Rural urban migration and government policies in low income countries', *International Labour Review* 118, 747–62.

Peek, P. and Standing, G. (1982a) 'State policies and labour migration', in P. Peek and G. Standing (eds.) *State Policies and Migration: Studies in Latin America and the Caribbean*, London, Croom Helm, pp. 1–34.

Peek, P. and Standing, G. (eds.) (1982b) *State Policies and Migration: Studies in Latin America and the Caribbean*, London, Croom Helm.

Perez-Sainz, J.P. and Zarembka, P. (1979) 'Accumulation and the state in Venezuelan industrialization, *Latin American Perspectives* 6, 5–29.

Phillips, A. (1977) 'The concept of development', *Review of African Political Economy*, 8, 7–20.

Phillips, L. (1987) 'Women, development, and the state in rural Ecuador', in C.D. Deere and M. Leon (eds.) *Rural Women and State Policy: Feminist Perspectives on Latin American Agricultural Development*, Boulder, Westview, pp. 105–23.

Platt, R.S. (1943) *Latin America: Countrysides and United Regions*, New York, McGraw-Hill.

Preston, D.A. (1980) 'Land tenure, rural emigration and rural development in highland Ecuador', in R.L. Singh and R.P.B. Singh (eds.) *Rural Habitat Transformation in World Frontiers*, Varanasi, NGSI Publications, pp. 182–91.

Preston, D.A. and Preston, R.A. (1983) 'Migration, education, and rural development: evidence from Ecuador', paper presented at the 1983 meeting of the Population Association of America, Pittsburgh.

Preston, D.A. and Taveras, G.A. (1976) 'Caracteristicas de la emigracion rural en la Sierra Ecuatoriana', *Revista Geografica* 84, 23–31.

Preston, D.A., Taveras, G.A., and Preston, R.A. (1979) 'Rural emigration and agricultural development in highland Ecuador: final report', Working Paper 238, Department of Geography, University of Leeds.

Programa Nacional de Regionalizacion Agraria (1979) *Diagnostico Socio Economico del Medio Rural Ecuatoriano. Documento A: Formacion de las Estructuras Agrarias en el Ecuador, Metodologia. Documento B: Las Zonas Socio Economicas Actuales Homogeneas de la Sierra. Documento C: Las Zonas Socio Economicas Actuales Homogeneas de la Costa. Documento D: Las Zonas Socio Economicas Actuales Homogeneas de la Region Amazonica Ecuatoriana y Conclusiones Generales a Nivel Nacional.* Quito, Instituto Latinoamericano de Investigaciones Sociales, Ministerio de Agricultura y Ganaderia.

Prothero, R.M. and Chapman, M. (eds.) (1985) *Circulation in Third World Countries*, London, Routledge and Kegan Paul.

Pryor, R. (1981) 'Population redistribution: policy formulation and implementation', in G.J. Demko and R.J. Fuchs (eds.) *Population Distribution Policies in Development Planning*, New York, United Nations, Department of Economic and Social Affairs, Population Studies 75, pp. 169–82.

Psacharopoulos, G. (1980) 'Returns to education: an updated international comparison', in T. King (ed.) *Education and Income*, Washington, World Bank, Staff Working Paper 402, pp. 73–109.

Psacharopoulos, G. and Hinchliffe, K. (1973) *Returns to Education: An International Comparison*, Amsterdam, Elsevier, San Francisco, Jossey Bass.

Psacharopoulos, G. and Woodhall, M. (1985) *Education for Development: An Analysis of Investment Choices*, Oxford and New York, Oxford University Press.

Pudup, M.B. (1988) 'Arguments within regional geography', *Progress in Human Geography* 12, 369–90.

Ramos, H. (1984) 'Agricultural credit situation', technical report, United States Agency for International Development, Ecuador Mission, Quito.

Recchini de Lattes, Z. (1983) *Dynamics of the Female Labour Force in Argentina*, Paris, United Nations Educational, Scientific, and Cultural Organization (UNESCO).

Recchini de Lattes, Z. and Wainerman, C.H. (1986) 'Unreliable account of women's work: evidence from Latin American census statistics', *Signs* 11, 740–50.

Reinhardt, N. (1987) 'Modernizing peasant agriculture: lessons from El Palmar, Colombia', *World Development* 15, 221–47.

Rempel, H. (1980) 'Determinants of rural-to-urban migration in Kenya', *International Institute for Applied Systems Analysis Reports* 2, 281–307.

Rempel, H. (1981) *Rural-Urban Labor Migration and Urban Unemploy-*

ment in Kenya, Laxenburg, International Institute for Applied Systems Analysis.

Renaud, B. (1981) *National Urbanization Policy in Developing Countries*, Oxford and New York, Oxford University Press.

Rhoda, R.E. (1979) *Development Activities and Rural-Urban Migration: Is It Possible to Keep Them Down on the Farm?*, Washington, Agency for International Development, Office of Urban Development.

Rhoda, R.E. (1983) 'Rural development and urban migration: can we keep them down on the farm?', *International Migration Review* 17, 34–64.

Richardson, H.W. (1976) 'Growth pole spillovers: the dynamics of backwash and spread', *Regional Studies* 10, 1–9.

Richardson, H.W. (1980) 'Polarization reversal in developing countries', *Papers of the Regional Science Association* 45, 67–85.

Richardson, H.W. (1981) 'National urban development strategies in developing countries', *Urban Studies* 18, 267–83.

Richardson, H.W. and Richardson, M. (1975) 'The relevance of growth center strategies to Latin America', *Economic Geography* 51, 163–78.

Riddell, J.B. (1970) *The Spatial Dynamics of Modernization in Sierra Leone: Structure, Diffusion, and Response*, Evanston, Northwestern University Press.

Riddell, J.B. (1976) 'Modernization in Sierra Leone', in C.J. Knight and J.L. Newman (eds.) *Contemporary Africa: Geography and Change*, Englewood Cliffs, Prentice-Hall, pp. 393–407.

Riddell, J.B. (1981) 'Beyond the description of spatial pattern: the process of proletarianization as a factor in population migration in West Africa', *Progress in Human Geography* 5, 370–92.

Roberts, B.R. (1977) 'Center and periphery in the development process: the case of Peru', in J. Abu-Lughod and R. Hay, Jr. (eds.) *Third World Urbanization*, Chicago, Maaroufa Press, pp. 176–93.

Roberts, B.R. (1978) *Cities of Peasants: The Political Economy of Urbanization in the Third World*, London, Edward Arnold, Beverly Hills, Sage.

Roberts, K.D. (1982) 'Agrarian structure and labor mobility in rural Mexico', *Population and Development Review* 8, 299–322.

Roberts, K.D. (1985) 'Household labour mobility in a modern agrarian economy: Mexico', in G. Standing (ed.) *Labour Circulation and the Labour Process*, London, Croom Helm, pp. 358–81.

Rodas, H. (1985) 'La migracion campesina en el Azuay', *Ecuador Debate* 8, 155–93.

Rodwin, L. (ed.) (1969) *Planning Urban Growth and Regional Development: The Experience of the Guayana Program of Venezuela*, Cambridge, MIT Press.

Rogers, E.M. (1969) *Modernization Among Peasants: The Impact of Communications*, New York, Holt, Rinehart, and Winston.

Rondinelli, D.A. (1983) *Secondary Cities in Developing Countries: Policies for Diffusing Urbanization*, Beverly Hills, Sage.

Rondinelli, D.A. and Evans, H. (1983) 'Integrated rural development planning: linking urban centers and rural areas in Bolivia', *World Development* 11, 31–54.

Rondinelli, D.A. and Ruddle, K. (1978) *Urbanization and Rural Development: A Spatial Policy for Equitable Growth*, New York, Praeger.

Roseberry, W. (1983) *Coffee and Capitalism in the Venezuelan Andes*, Austin, University of Texas Press.

Rostow, W.W. (1960) *The Stages of Economic Growth: A Non-Communist Manifesto*, Cambridge, Cambridge University Press.

Rudel, T.K. (1983) 'Roads, speculators, and colonization in the Ecuadorian Amazon', *Human Ecology* 11, 385–403.

Sabot, R.H. (1979) *Economic Development and Urban Migration: Tanzania, 1900–1971*, Oxford, Oxford University Press.

Saint, W.S. and Goldsmith, W.W. (1980) 'Cropping systems, structural change, and rural-urban migration in Brazil', *World Development* 8, 259–72.

Salvatore, D. (1981) *Internal Migration and Economic Development: A Theoretical and Empirical Study*, Washington, University Press of America.

Sayer, A. (1984) *Method in Social Science: A Realist Approach*, London, Hutchinson.

Sayer, A. (1985) 'Realism and geography', in R.J. Johnston (ed.) *The Future of Geography*, London and New York, Methuen, pp. 159–73.

Sayer, A. (1986) 'Industrial location on a world scale: the case of the semiconductor industry', in A.J. Scott and M. Storper (eds.) *Production, Work, Territory: The Geographical Anatomy of Industrial Capitalism*, London and Boston, Allen and Unwin, pp. 107–23.

Schlesinger, A.M., Jr. (1986) *The Cycles of American History*, Boston, Houghton Mifflin.

Schneider-Sliwa, R. and Brown, L.A. (1986) 'Rural-nonfarm employment and migration: evidence from Costa Rica', *Socio-Economic Planning Sciences* 20, 79–93.

Schodt, D.W. (1987) *Ecuador: An Andean Enigma*, Boulder, Westview.

Schultz, T.P. (1971) 'Rural-urban migration in Colombia', *Review of Economics and Statistics* 53, 157–63.

Schultz, T.P. (1982) 'Lifetime migration within educational strata in Venezuela: estimates of a logistic model', *Economic Development and Cultural Change* 30, 559–93.

Schultz, T.W. (1980) 'The economics of being poor', *Journal of Political Economy* 88, 639–51.

Scott, A.J. and Storper, M. (eds.) (1986) *Production, Work, Territory: The Geographical Anatomy of Industrial Capitalism*, London and Boston, Allen and Unwin.

Scott, I. (1982) *Urban and Spatial Development in Mexico*, Baltimore, Johns Hopkins University Press.

Segnini, I.S. (1974) 'Venezuela', *Encyclopedia Britannica, Fifteenth Edition*, New York, pp. 58–68.

Seligson, M.A. (1980) *Peasants of Costa Rica and the Development of Agrarian Capitalism*, Madison, University of Wisconsin Press.

Sheck, R.C., Brown, L.A., and Horton, F.E. (1971) 'Employment structure as an indicator of shifts in the space-economy: the case of the Central

American Common Market region', *Revista Geografica* 75, 49–72.

Sheppard, E. (1982) 'City size distributions and spatial economic change', *International Regional Science Review* 7, 127–51.

Short, J.F., Jr. (1971) *The Social Fabric of the Metropolis: Contributions of the Chicago School of Urban Sociology*, Chicago, University of Chicago Press.

Shrestha, N.R. (1987) 'Institutional policies and migration behavior: a selective review', *World Development* 15, 329–45.

Shryock, H.S. and Siegel, J.S. (1976) *The Methods and Materials of Demography*, New York, Academic Press.

Sigma One. (1985) 'Report on exchange rate and macro policy impacts on the agricultural sector', technical report, Ministerio de Agricultura y Ganaderia, Quito.

Silvers, A.L. and Crosson, P. (1983) 'Urban bound migration and rural investment: the case of Mexico', *Journal of Regional Science* 23, 33–47.

Simmons, R. and Ramos, H. (1985) 'Potato marketing in Ecuador', technical report, United States Agency for International Development, Ecuador Mission, Quito.

Simon, D. (1984) 'Third World colonial cities in context: conceptual and theoretical approaches with particular reference to Africa', *Progress in Human Geography* 8, 493–514.

Sjaastad, L.A. (1962) 'The costs and returns of human migration', *Journal of Political Economy* 70, 80–93.

Skeldon, R. (1977) 'The evolution of migration patterns during urbanization in Peru', *Geographical Review* 67, 394–411.

Skeldon, R. (1985) 'Circulation: a transition in mobility in Peru', in R.M. Prothero and M. Chapman (eds.) *Circulation in Third World Countries*. London, Routledge and Kegan Paul, pp. 100–20.

Smith, N. (1987) 'Dangers of the empirical turn: some comments on the CURS initiative', *Antipode* 19, 59–68.

Smith, W.R., Huh, W., and Demko, G.J. (1983) 'Population concentration in an urban system: Korea 1949–1980', *Urban Geography* 4, 63–79.

Soja, E. (1968) *The Geography of Modernization in Kenya*, Syracuse, Syracuse University Press.

Speare, A. (1971) 'A cost benefit analysis of rural to urban migration in Taiwan', *Population Studies* 25, 117–30.

Squire, L. (1979) *Labor Force, Employment, and Labor Markets in the Course of Economic Development*, Washington, World Bank, Staff Working Paper 336.

Standing, G. (1982) *Labour Force Participation and Development, Second Edition*, Geneva, International Labour Office.

Standing, G. (1985a) 'Circulation and the labour process', in G. Standing (ed.) *Labour Circulation and the Labour Process*, London, Croom Helm, pp. 1–45.

Standing, G. (ed.) (1985b) *Labour Circulation and the Labour Process*, London, Croom Helm.

Steinbeck, J. (1939) *The Grapes of Wrath*, New York, Viking Press.

Stohr, W.B. (1974) *Interurban Systems and Regional Economic Develop-*

ment, Washington, Association of American Geographers, Resource Paper 26.

Stohr, W.B. (1975) *Regional Development Experiences and Prospects in Latin America*, The Hague, Mouton.

Stohr, W.B. (1981) 'Development from below, the bottom-up and periphery-inward development paradigm', in W.B. Stohr and D.R.F. Taylor (eds.) *Development from Above or Below? The Dialectics of Regional Planning in Developing Countries*, New York, John Wiley, pp. 39–72.

Stohr, W.B. (1982) 'Structural characteristics of peripheral areas: the relevance of the stock-in-trade variables of regional science', *Papers of the Regional Science Association* 49, 71–84.

Stohr, W.B. and Taylor, D.R.F. (eds.) (1981) *Development from Above or Below? The Dialectics of Regional Planning in Developing Countries*, New York, John Wiley.

Swift, J. (1978) *Economic Development in Latin America*, New York, St. Martin's Press.

Swindell, K. (1979) 'Labour migration in underdeveloped countries: the case of Sub-Saharan Africa', *Progress in Human Geography* 3, 239–59.

Taaffe, E.J. (1974) 'The spatial view in context', *Annals of the Association of American Geographers* 64, 1–16.

Taaffe, E.J. (1985) 'Comments on regional geography', *Journal of Geography* 84, 96–7.

Taaffe, E.J., Morrill, R.L., and Gould, P.R. (1963) 'Transport expansion in underdeveloped countries: a comparative analysis', *Geographical Review* 53, 503–29.

Taglioretti, G. (1983) *Women and Work in Uruguay*, Paris, United Nations Educational, Scientific, and Cultural Organization (UNESCO).

Tang, A.M. and Worley, J.S. (eds.) (1988) *Why Does Overcrowded, Resource-poor East Asia Succeed – Lessons for the LDC's?*, Supplement Issue, *Economic Development and Cultural Change* 36.

Tata, R.J. and Schultz, R.R. (1988) 'World variation in human welfare: a new index of development status', *Annals of the Association of American Geographers* 78, 580–93.

Taylor, J.E. (1980) 'Peripheral capitalism and rural-urban migration: a study of population movements in Costa Rica', *Latin American Perspectives* 26, 75–90.

Thirsk, W.R. (1976) 'Price policy and agricultural development in Ecuador', technical report, Rice University.

Thomsen, M. (1969) *Living Poor: A Peace Corps Chronicle*, Seattle, University of Washington Press.

Tiano, S. (1986) 'Women and industrial development in Latin America', *Latin American Research Review* 21, 157–70.

Todaro, M.P. (1969) 'A model of labor migration and urban unemployment in less developed countries', *American Economic Review* 59, 138–48.

Todaro, M.P. (1971) 'Income expectations, rural-urban migration and employment in Africa', *International Labour Review* 104, 387–413.

Todaro, M.P. (1976) *Internal Migration in Developing Countries: A Review of Theory, Evidence, Methodology, and Research Priorities*, Geneva, International Labour Office.

Todaro, M.P. (1985) *Economic Development in the Third World, Third Edition*, New York and London, Longman.

Todaro, M.P. and Stilkind, J. (1981) *City Bias and Rural Neglect: The Dilemma of Urban Development*, New York, The Population Council.

Tolley, G.S., Thomas, V., and Wong, C.M. (1982) *Agricultural Price Policies and the Developing Countries*, Baltimore, Johns Hopkins University Press.

Townroe, P.M. and Keen, D. (1984) 'Polarization reversal in the state of Sao Paulo, Brazil', *Regional Studies* 18, 45–54.

Treiman, D.J. (1977) *Occupational Prestige in Comparative Perspective*, New York, Academic Press.

Urzua, R. (1981) 'Population redistribution mechanisms as related to various forms of development', in G.J. Demko and R.J. Fuchs (eds.) *Population Distribution Policies in Development Planning*, New York, United Nations, Department of Economic and Social Affairs, Population Studies 75, pp. 53–69.

Watts, M.J. and Bassett, T.J. (1986) 'Politics, the state and agrarian development: a comparative study of Nigeria and the Ivory Coast', *Political Geography Quarterly* 5, 103–25.

Weber, A.F. (1899) *The Growth of Cities in the Nineteenth Century: A Study in Statistics*, New York, MacMillan; reprinted by Cornell University Press, 1967.

Weeks, J. (1985) *Limits to Capitalist Development: The Industrialization of Peru, 1950–1980*, Boulder, Westview.

Weil, T.E., Black, J.K., Blutstein, H.I., McMorris, D.S., Mersereau, M.G., Munson, F.P., and Parachini, K.E. (1973) *Area Handbook for Ecuador*, Washington, United States Government Printing Office.

Weinstein, J.A. and McNulty, M.L. (1980) 'The interpenetration of modern and traditional structures: a spatial perspective', *Studies in Comparative International Development* 15, 45–61.

West, R.C. and Augelli, J.P. (1976) *Middle America: Its Lands and Peoples, Second Edition*, Englewood Cliffs, Prentice-Hall.

Wilber, C.K. and Jameson, K.P. (1988) 'Paradigms of economic development and beyond', in C.K. Wilber (ed.) *The Political Economy of Development and Underdevelopment, Fourth Edition*. New York, Random House, pp. 3–27.

Women and Geography Study Group of the IBG (1984) *Geography and Gender: An Introduction to Feminist Geography*, London, Hutchinson.

Wood, C.H. (1982) 'Equilibrium and historical-structural perspectives on migration', *International Migration Review* 16, 298–319.

World Bank (1979) *Ecuador: Development Problems and Prospects*, Washington, World Bank, Country Study Series.

World Bank (1980) *World Development Report, 1980*, Oxford and New York, Oxford University Press.

World Bank (1984) *Ecuador: An Agenda for Recovery and Sustained Growth*, Washington, World Bank, Country Study Series.

Wrigley, N. (1976) *Introduction to the Use of Logit Models in Geography*, Norwich, University of East Anglia, Geo Abstracts, CATMOG Series 10.

Yapa, L.S. (1977) 'Innovation diffusion and economic involution', *Antipode* 9, 20–9.

Yotopoulos, P.A. and Nugent, J.B. (1976) *Economics of Development: Empirical Investigations*, New York, Harper and Row.

Zelinsky, W. (1971) 'The hypothesis of the mobility transition', *Geographical Review* 61, 219–49.

Zelinsky, W., Monk, J., and Hanson, S. (1982) 'Women and geography: a review and prospectus', *Progress in Human Geography* 6, 317–66.

Zuvekas, C. (1975) 'Economic policy and economic development in Ecuador, 1950–1974', technical report, United States Agency for International Development, Ecuador Mission, Quito.

Index

apparently aspatial compared to explicitly spatial national policies, and regional change 129, 153, 159, 167–72, 180, 195–6, 201, 203

bridging conceptual and methodological perspectives 7–8, 197, 203–6

census data use (and substantive interpretation) 8, 83, 152, 205
Centro Latinoamericano de Demografia (CELADE, data source) 84, 87, 89, 108, 143, 214 note 7, 216 note 6, 219 note 11
circulation compared to migration 131–2, 144–53, 194
circulation defined 131–2
circulation factors: age 144; culture 146–7; drought 147; economic orientation 147, 149; educational attainment 144; farm size and tenancy 146; gender 144; land reform 134–6, 146, 150–4, 195–6; marital status 144; place characteristics related to development 144–9; proximity to economic activity 149, 151; temporary employment 134–5
Colombia, regional change 199
contextual characteristics: see place characteristics related to development
Costa Rica: aggregate migration flows in 54–78; banana production 31, 67, 69, 73, 75–6, 78–9, 190; cacao production 30, 69, 73, 76, 78–9, 190; cattle production 31, 46–7, 67, 73, 76, 78–9, 190; land reform–consolidation 31, 46, 67, 73–4, 76, 78–9, 190; Pan American highway expansion 31, 67, 74, 76, 78–9, 190; political

subdivisions 211 note 14; rural-directed migration in 46–7, 75–7, 212 note 22; rural-nonfarm activity 76; sugar cane production 30, 76, 78–9, 190; urban–rural core–periphery differences 30–1, 54, 59, 67, 74

development: multivariate compared to single variable indices 83–4, 98, 193; qualitative indicators 30–1, 79–80, also see place knowledge, sense of place; quantitative indices or measurement of 79–80, 83–7, also see place characteristics related to development
development and interest groups 207 note 4
development and migration, as geographic phenomenon 1–2
development models (frameworks, conceptualizations): and ground level reality or local conditions 4, 24–5, 34–7, 158–9, 191; and the dialectic of academic progress 36–7; need to rethink 3–4, 10, 34, 191; specific formulations: basic needs 15–17; complementarity between 18–20, 24–5, 28–9, 36–7, 47; core-periphery and growth center 13–15, 44–5, 48–50, 86–7, 181, 197, 200; historical-structural 45–6; human resource 15–17, 45, 107–8; Latin American school 21–4; modernization theory 13–15, 48–50, 83–4; orthodox/neoclassical compared to political economy approaches 10, 26–9; political economy 17–24, 45–7, 207 note 1; stages of growth 11–12, 48, 51; two-sector growth 13–15,

44–5, 48–50
spatial scales 24–6

development paradigm of migration 47–54, 61–2, 77–8, 98, 191, 213 note 2
donor-nation actions and development models 30; *also see* exogenous forces
donor-nation actions, occurrences reflecting: miscellaneous examples 25, 30, 32, 202; Pan American highway and transportation expansion in Costa Rica 31, 67, 74, 76, 78–9, 190

economic enterprise in: Colombia 35, 199; Ecuador 31, 34–5, 197–8; Kenya 36; Peru 35, 198–9; United States 33, 37–9; Yucatan 32
economic man 207 note 2
Ecuador: agrarian structure 132–6; agricultural credit 5, 128, 134–5, 157, 166, 169–70, 172–6, 180, 195–6; agricultural pricing 5, 128, 157, 166–9, 172–6, 180, 195–6; Costa region 129–30; description of country 129–36; elements of change locally 31; hacienda system 133–5, 150–1; import substitution industrialization 5, 99, 128, 130–1, 154, 157, 166–7, 169, 171, 173, 179–81, 191, 195, 197; land reform 5, 99, 128–38, 140, 143, 146, 149–54, 157, 159, 162–5, 179–81, 191, 194–7, 202, 212 note 19, 219 notes 8 and 13; monetary exchange 5, 128, 157, 170–6, 180, 195–6; Oriente or Amazon region 129–30; out-migration and out-circulation in 81–2, 135–6, 144–54; petroleum production 5, 130–1, 133, 135–6, 160, 164, 166–7, 171–2, 179–81, 197, 207 note 4; political sub-

divisions 213 note 3, 218 notes 1 and 3; shrimp production 131; Sierra region 129–30
exogenous forces: *also see* donor-nation actions, national policies, world markets (political and economic conditions)
exogenous forces: and regional change 17, 25, 26, 31, 34, 36, 157–9, 181, 195–203, 205; classification 202–3; examples 25–6, 195, 202; interplay with place characteristics and local articulation 4–6, 25, 79–80, 84, 99, 128–9, 132, 146–54, 159, 191–2, 195–6, 203, *also see* local conditions or ground level reality

gender differentials in labor market experiences 3, 106–27, 190–1, 196, 217 note 8, 217–18 note 11; and census data 217 note 7
gold production in Ecuador 197–9

human resources in development 15–17, 41, 44–5, 94, 107–8, 160, 198
hypothesis of the mobility transition 50–2, 213 note 2

idiographic research strategies, *see* nomothetic
interaction terms, interpretation of 93–4, 113–14

labor market experiences 79–80; 106–26
labor markets and the spatial division of labor 204
land reform, land consolidation: in Costa Rica 31, 46, 67, 73–4, 76, 78–9, 190; in Ecuador 5, 99, 128–38, 140, 143, 146, 149–54, 157, 159, 162–5, 179–81, 191, 194–7, 202, 212 note 19, 219 notes 8 and 13;

in Mexico 222 note 7
local conditions or ground level reality: *also see* place characteristics related to development; place knowledge, sense of place, place reality
local conditions or ground level reality: and development frameworks 24, 196–7; influence on migration 59–77; role in development 4–5; role in filtering or conditioning exogenous forces, *see* exogenous forces
locality studies or the new regional geography 202

migration: link with development 44–54, 78, 80–1, *also see* development paradigm of migration, migration factors; rural directed 46–7, 51–3, 75–7, 212 note 22
migration and development as geographic phenomenon 1–2
migration compared to circulation; *see* circulation compared to migration
migration factors accounting for aggregate flows: amenity levels 80; development milieu 47–54, 61–2, 74; distance 44, 52, 54, 57–72, 76–7; educational opportunities 52–3; employment opportunities 42–4, 48–9, 52–3, 57–72, 74, 80, 190; information flows and migration chain effects 42, 44, 48–50, 52–3, 61, 74, 77, 80, 190; land distribution and socioeconomic disparities 49–50, 77; movement costs 80; origin push 46–50, 52–4, 57–61, 73, 75, 210 note 5; population pressure 57–72, 76–7, 210 note 5; population size 57–72; rural and rural-nonfarm labor markets 76–7; social system characteristics 48, 50, 52–3; transportation and communications 48, 50, 52–3, 58; urban–rural differences 52–4, 57–72, 74; wage levels 42–4, 48–9, 57–72, 74, 76–7, 80
migration factors accounting for individual movements: age 80–2, 89–98, 144–6, 190; economic orientation 149–50, 152–3, 194; educational attainment 80–2, 89–98, 107, 125–6, 144–6, 190; employment status 80; family composition, migration experience, and socioeconomic characteristics 81–3; farm size and tenancy 149–50; gender 82, 89–98, 107, 125–6, 144–6, 190–1; information availability 96; land reform 146, 150–3, 191, 194; marital status, 81–2, 144–6; occupation 80, 82; personal resources 82–3, 96; place (or community) characteristics related to development 80–3, 89–98, 107, 125–6, 144–6, 149–50, 152–3, 190–1, 193–4; proximity to economic activity 107, 125–6 149, 151, 196; urban–rural differences 83, 89–98
migration models: conventional 42–4; human capital 43, 45, 209 note 1; labor force adjustment 42–3; migration chain 43–4, 209–10 notes 2 and 3; range of applicability 74–5; role of development therein 44–54, 61–2, 73–4, 77–8, 80–1, 98–9
modernization compared to paradigmatic-type development 37–9, 196–7, 207 note 5, *also see* pattern versus process in development occurrences

national policy and development models 29–30, 158–9, *also see* exogenous forces
national policy, examples:

from Costa Rica: land reform-consolidation 31, 46, 67, 73–4, 76, 78–9, 190; modernization of cattle production 31, 46–7, 67, 73, 76, 78–9, 190; transportation (and other infrastructure) 31, 67, 74, 76, 78–9, 190

from Ecuador: agricultural credit 5, 128, 134–5, 157, 166, 169–70, 172–6, 180, 195–6; agricultural pricing 5, 128, 157, 166–9, 172–6, 180, 195–6; import substitution industrialization 5, 99, 128, 130–1, 154, 157, 166–7, 169, 171, 173, 179–81, 191, 195, 197; land reform—consolidation 5, 99, 128–38, 140, 143, 146, 149–54, 157, 159, 162–5, 179–81, 191, 194–7, 202, 212 note 19, 219 notes 8 and 13; monetary exchange 5, 128, 157, 170–6, 180, 195–6

from Venezuela: dissemination of petroleum wealth benefits 107, 125, 196; economic decentralization 49, 87, 99, 125–6, 129, 153, 157, 190–1, 196; educational opportunity 99, 107, 115–16, 125–6, 129, 153, 157, 190–1, 196; regional development 87, 99, 107, 109, 125–6, 129, 157, 190–1, 196

other: agricultural credit 203, 212 note 23; agricultural pricing 203, 222 note 7; economic decentralization 29, 203, 213 note 23; education 29, 38, 203, 213 note 23; export-led growth—industrialization 25, 195, 202, 221 note 2; import substitution industrialization 17–18, 21, 25, 29–30, 195, 199, 202–3; land reform 202–3, 212 note 23, 222 note 7; monetary exchange 203; transportation (and other infrastructure) 29–30, 199, 203

national policy, occurrences reflecting:
in Costa Rica: cattle production 31, 46–7, 67, 73, 76, 78–9, 190; land reform—consolidation 31, 46, 67, 73–4, 76, 78–9, 190; Pan American highway and transportation expansion 31, 67, 74, 76, 78–9, 190

in Ecuador: bias toward imported rather than locally produced goods 166–7, 170–1; bias toward large agricultural producers rather than small 168–72, 207 note 4; bias toward production for export rather than domestic markets 166–9, 171–2; changes in employment patterns, land tenancy, land distribution, and structure of production 134–6, 146–51, 157; gold production 197–8; petroleum production 130–1, 164, 166–7, 171–2, 179–80, 207 note 4; spatial differentiation in impacts 131–2, 146–54, 157, 167–72, 179–81; urban bias 167, 169, 172, 207 note 4

in Venezuela: Ciudad Guayana and other regional development efforts 87, 99, 107, 109, 125–6, 129, 157, 190–1, 196; decentralization and location of employment opportunities 3, 49, 87, 99, 107, 109, 119–20, 122–3, 125–6, 129, 153, 157, 190–1, 196; decentralization of educational opportunities 3, 99, 107, 115–16, 119, 125–6, 129, 153, 157 190–1, 196; dissemination of petroleum wealth benefits 107, 125, 196; spatial disparity between employment opportunities and educated work force 3, 107, 119, 125–6, 190–1, 196; other examples: out-migration 212–13 note 23

national policy, spatial compared to apparently aspatial national policies: *see* apparently aspatial

nomothetic versus idiographic research strategies 7, 39, 41–2, 74–5, 192, 202, 206

North American–Western European compared to Third World development 37–9, 208 note 9

odyssey of research 6–9

panama hat production in Ecuador 198

paradigmatic views of development 3–4, 22–3, 26–39, 158–9, 181, 200, 204

pattern versus process in development occurrences 30, 181, 197–8, 200

Peru, regional change 35, 198–9, 208 note 8

Philippines: intention to move 82–3; out-migration and out-circulation in 82

physical quality of life index (PQLI) 214 note 4

place characteristics related to development: *also see* circulation factors, development, migration factors

place characteristics related to development 1–6, 8, 79–81, 97–8, 106, 152, 190; effect on equity and access to opportunity 106, 117–19, 124–5, 128, 208–9 note 10; mechanisms underlying 78, 98–9, 106–7, 112, 125–6, 128, 158–9, 191, 195–7; use of scaling procedures to derive indices 79–80, 83–4, 98, 193, 214 note 4

place characteristics related to

development, details on indices:

Ecuador agrarian settings: MODERNITY 138–40, 143; PRODUCTION ORIENTATION 137, 142–3; SIZE 137–8, 140; TIME 137, 140, 142

Ecuador regional change groupings: COMMERCIAL AGRICULTURE CANTONES 164–6; DOMESTIC AGRICULTURE CANTONES 166; MAJOR URBAN CANTONES 163–4; REGIONAL CENTER CANTONES 164

Ecuador regional change indices on which regional change groupings are based 159–63

Venezuela national development surface: 86–8; composite development index 214–15 note 10; PRESSURE 86, 193; STRUCTURE 84–6, 193

place differences and generalization 4, 153–4, 194

place knowledge, sense of place, place reality 8–9, 23, 25–6, 34, 78, 132, 152–4, 158, 192, 203–4, 222 note 8; role in research 30, 34, 39, 41, 78, 154, 192–4, 200–6, 221–2 note 4

polarization reversal 14, 16, 45, 119, 181, 197, 210 note 4

qualitative knowledge and the interpretation of quantitative findings 8–9, 154, 195–6, 205, *also see* place knowledge

regional change including role of exogenous forces and approach to studying 3–6, 24–5, 32, 34–7, 73, 78, 80–1, 153–4, 157–9, 176–7, 179–81, 191–2, 194–206; in contrast to development 25, 37–9; *also see* apparently aspatial compared to explicitly spatial national policies and regional change, Colombia, Ecuador, Peru, Yucatan Peninsula

regional geography 7, 75, 152, 202, 204, 206, 222 note 8

research techniques: discriminant analysis 176–80; discriminant iterations grouping algorithm combined with principle components analysis to derive indices of regional change 159–66; dummy (1/0) dependent variable 89, 112–13, 144, 219–20 note 15; interaction terms 89–90, 93–4, 113–14; linking contextual or place characteristics, individual attributes, and individual behavior 87, 89–90, 112–14, 143–4; logistic regression 89, 144, 215 note 12; migration rate dependent variable 56–8, 211 notes 12, 13, 15; principle components analysis to derive indices of place characteristics 83–8, 136–43, 159–63; regression 55–8, 112–14; regression with spatially varying parameters 55–8, 62–73, 211 note 11, 212 note 20; representing structural conditions relevant to policy impacts, variable selection 172–6; simultaneous equations and two-stage least squares 47, 210 note 10; standard ellipse 109–10

scaling procedures to define development indices 79–80, 83–4, 98, 193, 214 note 4; also see place characteristics related to development, details on indices

social categories, definition of term 106, 215–16 note 1

spatial compared to apparently aspatial national policies: see apparently aspatial

spatial division of labor 204, 208 note 7

structural tension 107

summary statements of: book 1–9, 190–2; Chapter 2 10, 39–40; Chapter 3 41–2, 72–4, 77–8; Chapter 4 79–80, 96–9; Chapter 5 106–7, 123–6; Chapter 6 128–9, 151–4; Chapter 7 157, 180–1; Chapter 8 200–6

summed b-coefficients, see interaction terms

Third World landscapes, forces of change, and paradigmatic views of development 30–7, also see paradigmatic views of development

transportation system development, ideal–typical sequence 18–19

unevenness of development 1, 10, 38, 106, 158–9, 181, 191–2

United States development 33, 37–9, 207–8 note 6, 208 note 9

urban bias, orientation, change 5, 31, 36, 46–7, 74, 86–7, 98–9, 158–9, 167, 169, 172, 179–81, 197, 203, 207 note 4, 210 note 5, 211–12 note 16, 215 note 10, 216 note 2; also see Venezuela, urban landscape

Venezuela: Ciudad Guayana and other regional development efforts 87, 99, 107, 109, 125–6, 129, 157, 190–1, 196; decentralization and location of economic activity 3, 49, 87, 99, 107, 109, 119–20, 122–3, 125–6, 129, 153, 157, 190–1, 196; decentralization of educational opportunities 3, 97, 99, 107, 115–16, 119, 125–6, 129, 153, 157, 190–1, 196; development indices 62, 83–7, 100–5, 109–12; development surface 86–8; dissemination of petroleum wealth benefits 107, 125, 196; educational attainment 3, 106–17 123–4; employment opportunities 3, 49, 93, 97, 107, 119–20, 122–3, 196; gender differentials in labor market experiences 3, 93, 97, 106–27, 190–1, 196, 217 notes 7 and 8, 217–18 note 11; gender differentials in out-migration 3, 87–98, 107, 190–1; human resources supplied by core areas to periphery 16, 94; labor force participation 106–14, 117–19, 123–4; labor market experiences 3, 79, 106–27; migration 2, 3, 43, 61–2, 75, 79, 131; out-migration in 3, 79–81, 87–99, 127; political subdivisions 214 note 6; spatial disparities between employment opportunities and educated work force 3, 107, 119, 125–6, 190–1, 196; urban landscape 49, 84–91, 93, 98–9, 100–5, 107, 125; wages 106–14, 119–24

West African urban system, impact of colonization 18–19
world markets (political and economic conditions), *also see* exogenous forces
world markets (political and economic conditions), occurrences reflecting:
  in Costa Rica: banana production 31, 67, 69, 73, 75–6, 78–9, 190; cacao production 30, 69, 73, 76, 78–9, 190; export crop production 74; sugar cane production 30, 76, 78–9, 190;
  in Ecuador: development of Oriente 130; export crop production 130–1, 134; gold

production 197–9; national policies 166–7, 170–2; panama hat production 198; patterns of regional change 179–81, 197; petroleum production and related activities 5, 130–1, 133, 135–6, 154, 160, 164, 166–7, 171–2, 179–81, 197, 207 note 4; shift from feudal to commodified labor and production structures 128, 134–6, 149–53, 177, 194, 202; shrimp production 131
other examples 25, 34, 38, 78, 195, 202; in Colombia 199, 201; in Latin America 17; in Peru 35; in Yucatan 32, 199; *also see* spatial division of labor

Yucatan Peninsula 32